Stem Cell Biology and Regenerative Medicine

AF787839

Series Editor
Kursad Turksen, Ph.D.
kturksen@ohri.ca

For further volumes:
http://www.springer.com/series/7896

Michal Amit • Joseph Itskovitz-Eldor

Atlas of Human Pluripotent Stem Cells

Derivation and Culturing

With contributions by Ilana Laevsky, BA
and Atara Novak, MSc

Humana Press

Michal Amit, PhD
Department of Obstetrics and Gynecology
Rambam Health Care Campus
Stem Cell Research Center
Rapapport Faculty of Medicine
Technion – Israel Insitute of Technology
Haifa, Israel
mamit@tx.technion.ac.il

Joseph Itskovitz-Eldor, MD DSc
Department of Obstetrics and Gynecology
Rambam Health Care Campus
Stem Cell Research Center
Rapapport Faculty of Medicine
Technion – Israel Insitute of Technology
Haifa, Israel
itskovitz@rambam.health.gov.il

ISBN 978-1-4939-5786-6 ISBN 978-1-61779-548-0 (eBook)
DOI 10.1007/978-1-61779-548-0
Springer New York Dordrecht Heidelberg London

© Springer Science+Business Media, LLC 2012
Softcover re-print of the Hardcover 1st edition 2012
All rights reserved. This work may not be translated or copied in whole or in part without the written permission of the publisher (Humana Press, c/o Springer Science+Business Media, LLC, 233 Spring Street, New York, NY 10013, USA), except for brief excerpts in connection with reviews or scholarly analysis. Use in connection with any form of information storage and retrieval, electronic adaptation, computer software, or by similar or dissimilar methodology now known or hereafter developed is forbidden.
The use in this publication of trade names, trademarks, service marks, and similar terms, even if they are not identified as such, is not to be taken as an expression of opinion as to whether or not they are subject to proprietary rights.

Printed on acid-free paper

Humana Press is part of Springer Science+Business Media (www.springer.com)

Preface

We are very pleased to present this *Atlas on Human Pluripotent Stem Cells—Derivation and Culturing*, summarizing 12 years of our team's experience, skill and knowledge in the derivation, culture and expansion of human embryonic stem cells, and more recently, of human induced pluripotent stem cells. The exploration and realization of the incredible potential of these pluripotent stem cells for studying the fundamentals of early human development and organogenesis, for cell-based therapy and for disease modeling in the dish are a key focus of current biomedical research. Although murine embryonic stem cells have been used in research for 4 decades, techniques established for their growth have proven less amenable for long-term culture, expansion and manipulation of human pluripotent stem cells.

The culmination of two events has enabled the production of human embryonic stem cells: the birth of the first IVF test-tube baby in 1978 (Steptoe and Edwards 1978) and the discovery in 1981 by Martin (1981) and Evans and Kaufman (1981) of mouse embryonic stem cells. Later, development of sequential media in the mid-1990s enabled the growth of fertilized oocytes to viable blastocysts, from which the inner cell mass was extracted and human embryonic stem cells derived. These breakthroughs paved the way to the derivation of the first five human embryonic stem cell lines by Thomson et al. in Madison, Wisconsin, 1998 (Thomson et al. 1998). More recently, Yamanaka and team enthused the scientific community with their publication on the reprogramming of adult skin fibroblasts into induced pluripotent stem cells (Takahashi and Yamanaka, 2006).

To realize the full potential of embryonic and induced pluripotent stem cells, technologies, and especially those related to stem cell-based therapies, must achieve controlled cell growth in defined conditions for prolonged time periods, while maintaining cell stability, i.e., minimal genetic abnormalities, pluripotency and differentiation potential. For stem cell-based therapies and screening, robust production of cells in controlled dynamic cultures (bioreactors) is required. To that end, methods for expansion of pluripotent stem cells in non-adherent conditions, i.e., in suspension, are emerging.

This atlas provides up-to-date techniques that will be useful to those currently active in basic as well as translational research in the field of embryonic and induced pluripotent stem cells. It commences with practical aspects of the derivation and growth of human embryonic stem cells from inner cell mass blastocyst stage embryos. Three chapters in this volume deal with cell culture techniques, presenting the protocols and morphology of cells cultured on mouse embryonic fibroblasts and on foreskin fibroblasts, and the culturing of cells in feeder-free conditions. Taken together, the information provided in these chapters will enable the culture of pluripotent stem cells in defined conditions that are animal product-free, serum-free and feeder-free. The subsequent chapter describes the transformation of cell growth from adhesion to non-adhesion cultures, laying the foundation for the development of a system for robust therapeutic and industrial modalities.

The pluripotency and differentiation potential of human pluripotent stem cells are examined and described in the two chapters that focus on the differentiation of the cells into embryoid bodies in vitro and teratoma formation in vivo. The differentiation by immunostaining of undifferentiated and early differentiated human pluripotent stem cells is demonstrated in the subsequent chapter. Karyotype stability of human pluripotent stem cells is sensitive to growth conditions and to the manner in which cells are handled. This important issue is discussed in another chapter describing the common principles of karyotyping and fluorescent in situ hybridization (FISH) methods as they apply to the field of pluripotent stem cells.

Induced pluripotent stem cells attract great interest for their potential in understanding the basics of cell reprogramming, personalized medicine and disease modeling—a topic that concludes this atlas. In this last chapter the method for the derivation of human induced pluripotent stem cells from hair follicle keratinocytes is described.

We hope that this concise yet comprehensive atlas becomes a reference and an encyclopedia for young as well as established researchers, students and other individuals, involved in the field of stem cells. It is also our hope that the methods, descriptions and images provided in this atlas facilitate the realization of the enormous potential of human pluripotent stem cells and shorten the path from the bench to the bedside.

We are grateful for the generous support of the Technion's Stem Cell Center by the Sohnis and Forman Families. We thank Ilana Laevsky and Atara Novak for their valuable contributions. The cooperation and enthusiasm shown by the staff members at Springer who were involved in the accomplishment of this project are greatly appreciated. Special thanks are extended to all the members of our laboratory who contributed during the past decade to the research presented in this book.

Haifa, Israel

Michal Amit, PhD
Joseph Itskovitz-Eldor, MD DSc

References

Evans MJ, Kaufman MH (1981) Establishment in culture of pluripotential cells from mouse embryos. Nature 292(5819):154–156

Martin GR (1981) Isolation of a pluripotent cell line from early mouse embryos cultured in medium conditioned by teratocarcinoma stem cells. Proc Natl Acad Sci U S A 78(12):7634–7638

Steptoe PC, Edwards RG (1978) Birth after the reimplantation of a human embryo. Lancet 2(8085):366

Takahashi K, Yamanaka S (2006) Induction of pluripotent stem cells from mouse embryonic and adult fibroblast cultures by defined factors. Cell 126(4):663–676

Thomson JA, Itskovitz-Eldor J, Shapiro SS, Waknitz MA, Swiergiel JJ, Marshall VS, Jones JM (1998) Embryonic stem cell lines derived from human blastocysts. Science 282(5391): 1145–1147

Contents

List of Abbreviations

bFGF	Basic fibroblast growth factor
Bio	Glycogen synthase kinase-3 specific inhibitor
C.R.A.	Chromosome resolution additive
DAPI	4′,6-diamidino-2-phenylindole
DMEM	Dulbecco's modified Eagle's medium
DMSO	Dimethyl sulfoxide
DNA	Deoxyribonucleic acid
EBs	Embryoid bodies
EGF	Epidermal growth factor
FBS	Fetal bovine serum
FISH	Fluorescent in situ hybridization
FITC	Fluorescein isothiocyanate
GSK-3	Glycogen synthase kinase-3
H&E	Hematoxylin and eosin
hESCs	Human embryonic stem cells
HFF	Foreskin fibroblasts
HRP	Horseradish peroxidase
ICM	Inner cell mass
ICR	Imprinting control region mice
IHC	Immunohistochemistry
IF	Immunofluorescence
IRS	Inner root sheath
iPSCs	Induced pluripotent stem cells
ISCN	International system for human cytogenetic nomenclature
IVF	In vitro fertilization
KO	Knockout

LIF	Leukemia inhibitory factor
MEFs	Mouse embryonic fibroblasts
NGS	Normal goat serum
NOR	Nucleolus organizer regions
ORS	Outer root sheath
PBS	Phosphate buffer saline
PGD	Pre-implantation genetic diagnosis
ROCK	Rho kinase inhibitor Y-27632
SCID	Severe combined immunodeficiency
SR	Serum replacement
STR	Stirred tank bioreactor
TRDF	Technion Research and Development Foundation
TGFβ1	Transforming growth factor beta 1
VEGF-A165	Vascular endothelial growth factor A (VEGF-A165)
ZP	Zona pellucida

Chapter 1
Methods for the Derivation of Human Embryonic Stem Cell Lines

Abstract Human embryonic stem cells (hESCs) are pluripotent cells derived from the inner cell mass (ICM) of the developing embryo. They have tremendous potential for the research of early human development, differentiation processes, and teratology, as well as for industrial and clinical purposes, such as drug screening and cell-based therapy. Since hESCs were first derived by Thomson and his colleagues in 1998, considerable effort has been invested to improve methods of isolating new lines, in defined conditions, and to increase success rates. This chapter discusses the most commonly used methods for deriving hESC lines, including immunosurgery and the whole embryo approach.

1.1 Introduction

Human embryonic stem cell (hESC) lines, like those from other species, are pluripotent cell lines derived from the inner cell mass (ICM) of the developing embryo. Due to their exceptional capability of proliferating indefinitely as undifferentiated cells when cultured in appropriate conditions, and of sustaining a normal karyotype, hESCs may have broad applications for industrial uses; clinical purposes, namely, cell-based therapy; and research of early human development, differentiation mechanisms, and lineage commitment.

The first ESC lines were derived from mouse embryos in 1981 (Evans and Kaufman 1981; Martin 1981). Bongso and colleagues later demonstrated that ESC-like cells can be isolated from human embryos. They described the isolation of ICM cells from human blastocysts and the culturing of these cells for two passages, while expressing alkaline phosphate activity and demonstrating ESC-like morphology (Bongso et al. 1994). In 1998, Thomson and his colleagues presented the first five hESC lines (Thomson et al. 1998). The lack of a source of high-quality human blastocysts accounts in large part for the long time lapse between the derivation of ESC lines from mouse embryo and the derivation of human lines using basically the same culturing techniques.

M. Amit and J. Itskovitz-Eldor, *Atlas of Human Pluripotent Stem Cells: Derivation and Culturing*, Stem Cell Biology and Regenerative Medicine, DOI 10.1007/978-1-61779-548-0_1, © Springer Science+Business Media, LLC 2012

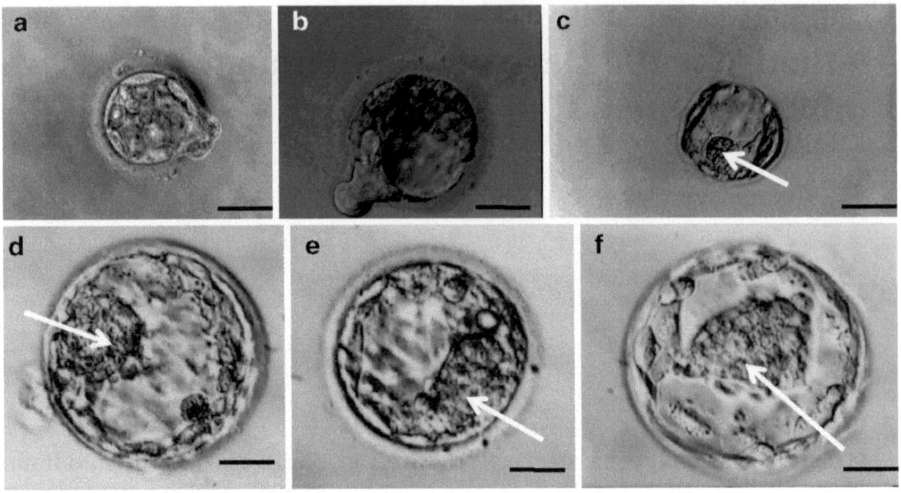

Fig. 1.1 Examples of human blastocyst morphology. (**a**) Five-day-old embryo in which the trophoplast extends through a hole drilled in the zona pellucida (ZP). No inner cell mass (ICM) can be recognized. (**b**) Six-day-old pseudo-blastocyst in which the trophoplast extends through a hole drilled for pre-implantation genetic diagnosis (PGD). No ICM can be recognized. (**c–f**) Six-day-old blastocysts with clear ICM. ICM is marked with *white arrows*. (**a–c**) Bar 50 μM, (**d–f**) Bar 30 μM

The first successful in vitro fertilization (IVF), based on retrieval of a single embryo from an oocyte developed in a spontaneous menstrual cycle, was reported in 1978 by the pioneering study of Steptoe and Edwards (1978). The technique of controlled ovarian hyperstimulation has since achieved several retrieved oocytes per cycle. In addition, the in vitro culturing of human embryos to the blastocyst stage has progressed greatly. The main source of human embryos for research purposes has thus become surplus IVF embryos (Thomson et al. 1998; Reubinoff et al. 2000; Amit and Itskovitz 2002; Cowan et al. 2004), as well as discarded low-quality embryos (Mitalipova et al. 2003; Lerou et al. 2008), abnormally fertilized zygotes, and discarded genetically abnormal embryos after pre-implantation genetic diagnosis (PGD) (Verlinsky et al. 2005; Mateizel et al. 2006; Amit and Itskovitz-Eldor, unpublished data). Only one documented study used embryos that were produced specifically for research purposes (Lanzendorf et al. 2001). Examples of human blastocyst morphology are illustrated in Fig. 1.1.

Similar to the methods used for the derivation of mouse ESCs, most hESC lines have been derived using supporting layers, mainly of mouse embryonic fibroblasts (MEFs) (Thomson et al. 1998; Reubinoff et al. 2000; Amit and Itskovitz-Eldor 2002; Cowan et al. 2004; Verlinsky et al. 2005). Klimanskaya and his colleagues first reported the feeder layer-free isolation of hESC lines. Using serum-free medium and MEF-produced matrix, they successfully generated six new hESC lines that exhibited ESC features after prolonged culture (Klimanskaya et al. 2005). This pioneering study

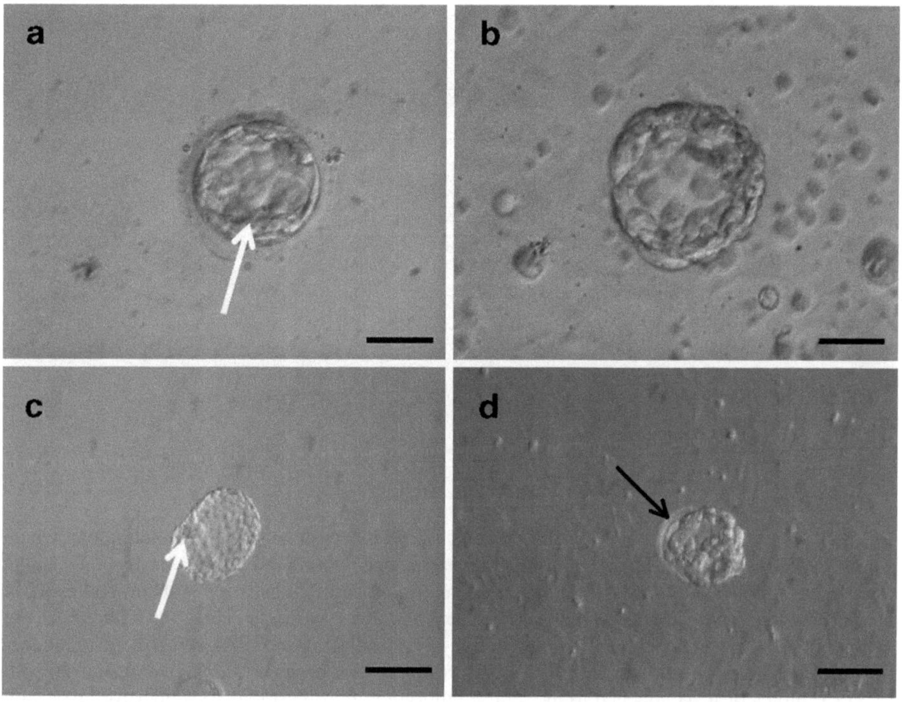

Fig. 1.2 Zona Pellucida removal with Tyrode's acid. (**a**) Six-day-old embryo with notable ICM (marked with *white arrow*) before ZP removal. (**b**) The same embryo after ZP removal. The ICM cannot be clearly distinguished. (**c**) Six-day-old blastocysts after ZP removal, with the ICM location clearly visible (marked with *white arrow*). (**d**) Six-day-old blastocyst after ZP was removed. A residue of the ZP is noted (*black arrow*). The ZP residue might affect the capability of the embryo to attach to the feeder layer. The ZP piece should be removed mechanically (enzymes or Tyrode's acid might harm the exposed embryo). (**a**, **b**) Bar 40 μM, (**c**, **d**) Bar 70 μM

demonstrates the feasibility of feeder layer-free derivation of hESCs. A recent publication by Ludwig and colleagues reported the derivation of two new hESC lines using a defined serum-free and animal-free medium, and feeder layer-free culture conditions (Ludwig et al. 2006). The matrix used consisted of human collagen, fibronectin, and laminin. The two newly derived cell lines sustained most hESC features after several months of continuous culture. However, both lines were reported to harbor karyotype abnormalities. It has yet to be determined whether the embryos were originally defected or whether the abnormalities detected are due to the method implemented.

ESC lines are traditionally isolated from blastocysts using immunosurgery, a straightforward method developed by Solter and Knowles during the 1970s for separating between the trophoectoderm layer and the ICM (Solter and Knowles 1975). This achieved, and following removal of the zona pellucida (ZP) (Fig. 1.2), the

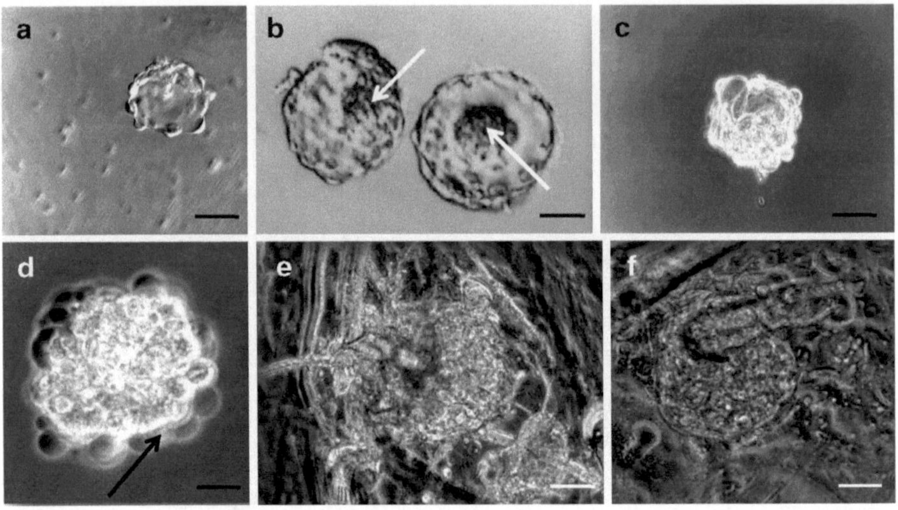

Fig. 1.3 Immunosurgery method. (**a**) Exposed embryo during incubation with antihuman whole serum antibodies. The ICM location is not clear; the trophectoderm morphology is distinguishable, as elongated cells. Bar 60 μM. (**b**) Two exposed embryos during incubation with antibodies; ICM is clearly noted (*white arrows*). Bar 30 μM. (**c**) Exposed embryo during incubation with guinea pig complement. The blastocoele is still notable. Bar 45 μM. (**d**) Exposed embryo at the end of incubation with guinea pig complement; ghost cells after lysis can be seen (*black arrow*). Bar 20 μM. (**e, f**) Examples of ICM attached to the mouse embryonic fibroblast (MEF) supportive layer 24 h post-plating. Bar 60 μM

embryo is exposed to antihuman whole serum antibodies, which attach to any human cell. Cell–cell connections within the outer layer of the trophoblast prevent penetration of the antibodies into the blastocyst, thus leaving the ICM cells intact. Incubation with guinea pig complement-containing medium then lyses all antibody-marked cells. The intact ICM is further rinsed and cultured with mitotically inactivated MEFs or an alternative feeder layer that supports hESC culture. The ICM can also be isolated mechanically by removing the trophectoderm layer, using either 25–27 gauge needles or pulled Puster pipettes under a dissecting microscope. As with immunosurgery, the isolated ICM is then cultured with a supporting layer. The steps of immunosurgery are depicted in Fig. 1.3. Following immunosurgery, ICM cells are cultured and allowed to proliferate, with MEFs serving as supporting layers. Although ICM cells can be split using trypsin or other enzymes (Cowan et al. 2004), higher survival rates may be achieved from mechanical splitting methods. The morphology of the colonies resulting from ICM growth and mechanical splitting is illustrated in Figs. 1.4–1.7.

Alternatively, ESC lines can be derived without the removal of the trophectoderm, by plating exposed embryos as a whole with a supporting layer, as illustrated in Figs. 1.8–1.10. The blastocyst, which is attached to the feeder layer, continues to grow with the surrounding trophoblast. When the ICM reaches sufficient size, it is selectively removed and propagated. ICM outgrowth within blastocysts plated

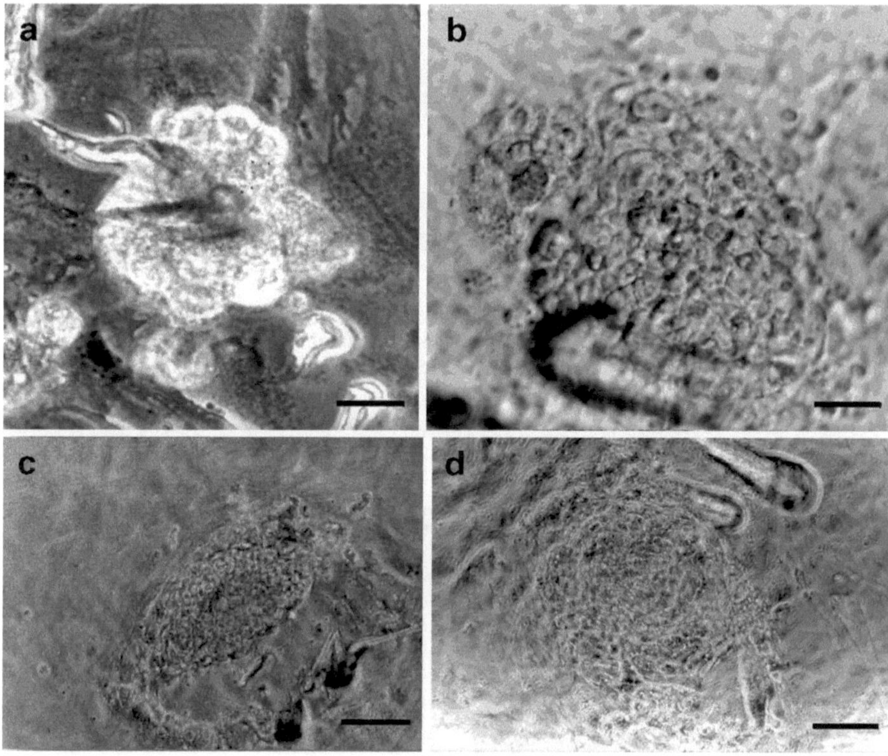

Fig. 1.4 Plated ICM. (**a**) ICM plated on MEFs 24 h post-plating. The ICM still resembles clumps of cells with no colony formation. Bar 50 μM. (**b–d**) ICM plated on MEFs 3 days post-plating. A colony was formed containing small round cells. Bar 20 μM, 90 μM for (**b**) and for (**c**, **d**), respectively

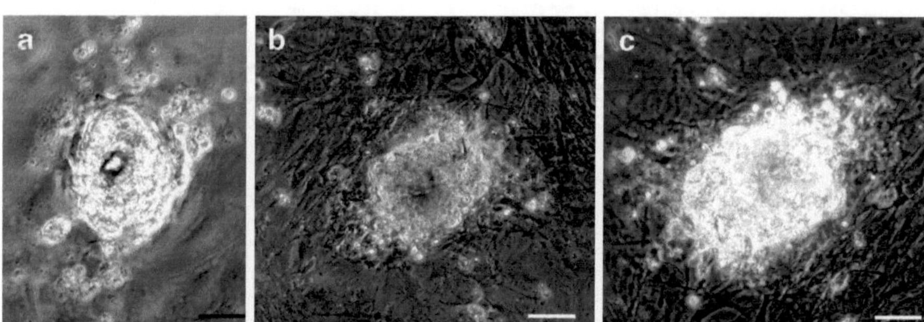

Fig. 1.5 Mechanical splitting. Examples of colony morphology 24 h post-mechanical splitting of the ICM and plating on MEFs. The cells are at passage 2. (**a**) Clear borders between MEFs and embryonic stem cells (ESCs) were formed, while in (**b**) and (**c**) the colonies were still disorganized. Bar 60 μM

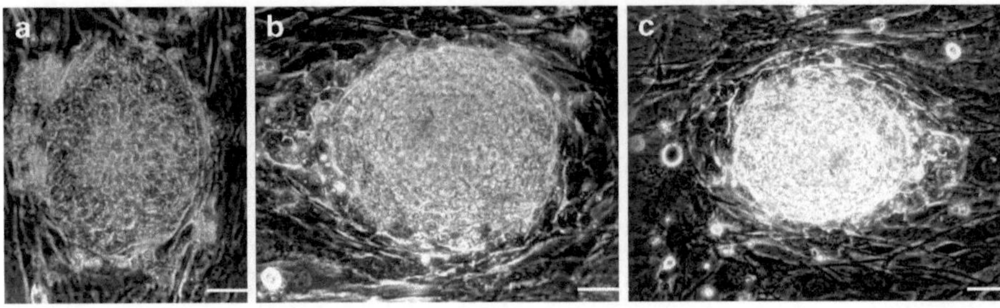

Fig. 1.6 Passage 2. Examples of colony morphology 3 days post-mechanical splitting of the ICM and plating on MEFs. The cells are at passage 2 post-derivation. All examples (**a–c**) demonstrate colony formation with borders between MEFs and ESCs, and outgrowth. In (**a**), the spaces between cells can be seen. Bar 60 μM

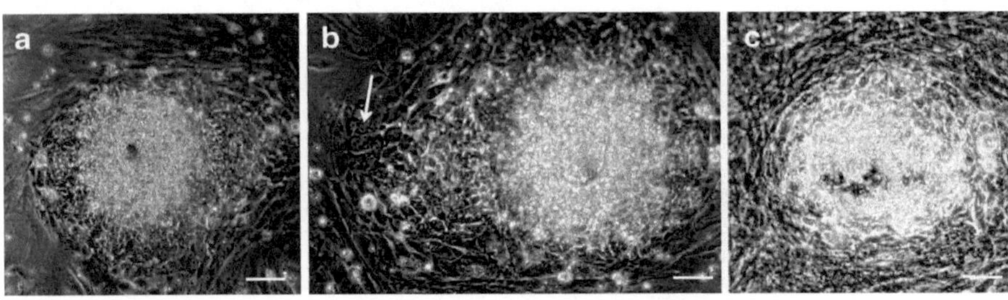

Fig. 1.7 Passage 2. Examples of colony morphology 4 days post-mechanical splitting of the ICM and plating on MEFs. The cells are at passage 2 post-derivation. All examples (**a–c**) demonstrate colony formation, outgrowth, and ESC typical morphology of small cells with large nuclei, notable nucleoli (*white arrow*), and spaces between cells. Bar 90 μM

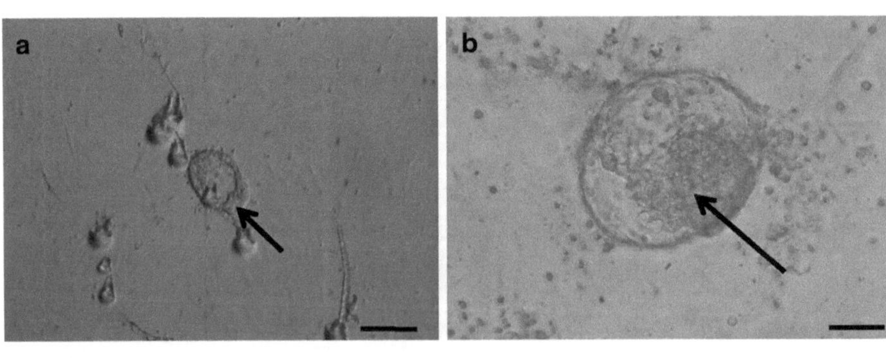

Fig. 1.8 Whole embryo approach for derivation of ESC lines. Exposed embryos plated on MEFs 12 h post-plating. In both embryos (**a**) and (**b**), the ICM is still recognized (*black arrows*) within the attached embryo, which still contains a cavity. Bar 60 μM and 40 μM for (**a**) and (**b**), respectively

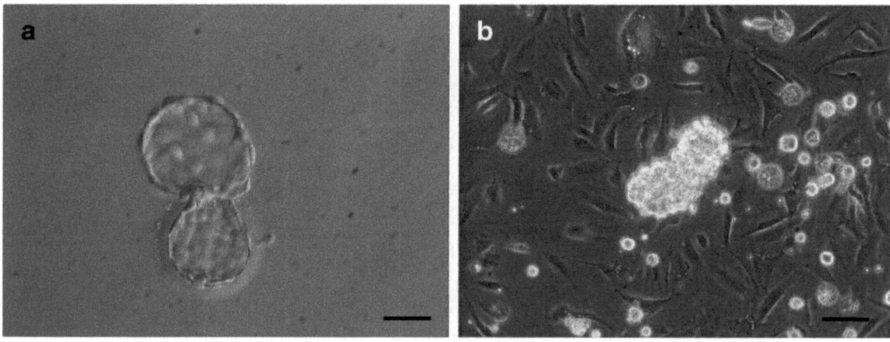

Fig. 1.9 Whole embryo approach for derivation of ESC lines. (**a**) Embryo before ZP removal. Due to the drill in ZP, a large part of the embryo hatched from the ZP. Therefore, the ZP was removed mechanically to avoid damage to the embryo that would be incurred by the use of Tyrode's acid or enzyme. (**b**) The exposed embryo plated on MEFs still has the same shape as when covered by the ZP. Bar 50 μM and 80 μM for (**a**) and (**b**), respectively

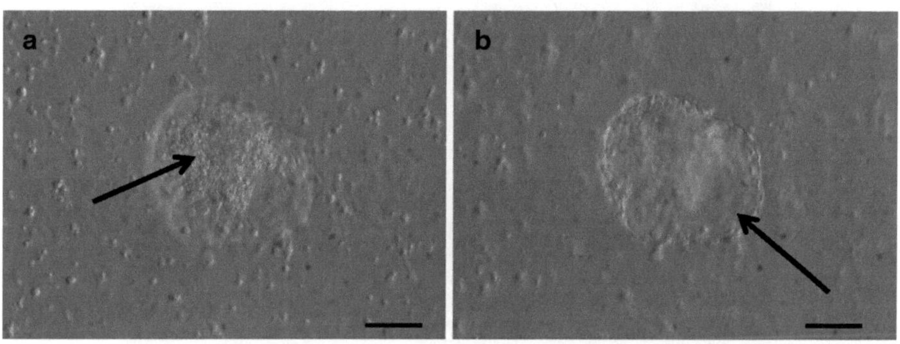

Fig. 1.10 Whole embryo approach for derivation of ESC lines—partial embryo attachment. Exposed embryo plated on MEFs 12 h post-plating. The embryo is only partly attached. The ICM cannot be located. (**a**) Close-up of the attached part of the embryo (*black arrow*). (**b**) Focus on the unattached trophoblasts (*black arrow*). Bar 50 μM

whole is demonstrated in Fig. 1.11. However, success rates in deriving hESC lines are lower with the whole embryo approach, and some ICM colonies differentiate (Figs. 1.12 and 1.13). Nevertheless, an advantage of the whole embryo approach is isolation of hESC lines without the use of animal products (antibodies). Figure 1.14 demonstrates derivation of the ESC line, CL1, using animal-free medium (NutriStem™, Biological Industries Ltd), human foreskin fibroblasts as feeder layers, and the whole embryo approach as the derivation method (Amit et al., unpublished data).

The increasing number of hESC lines attests to the fact that their isolation is a reproducible procedure with reasonable success rates.

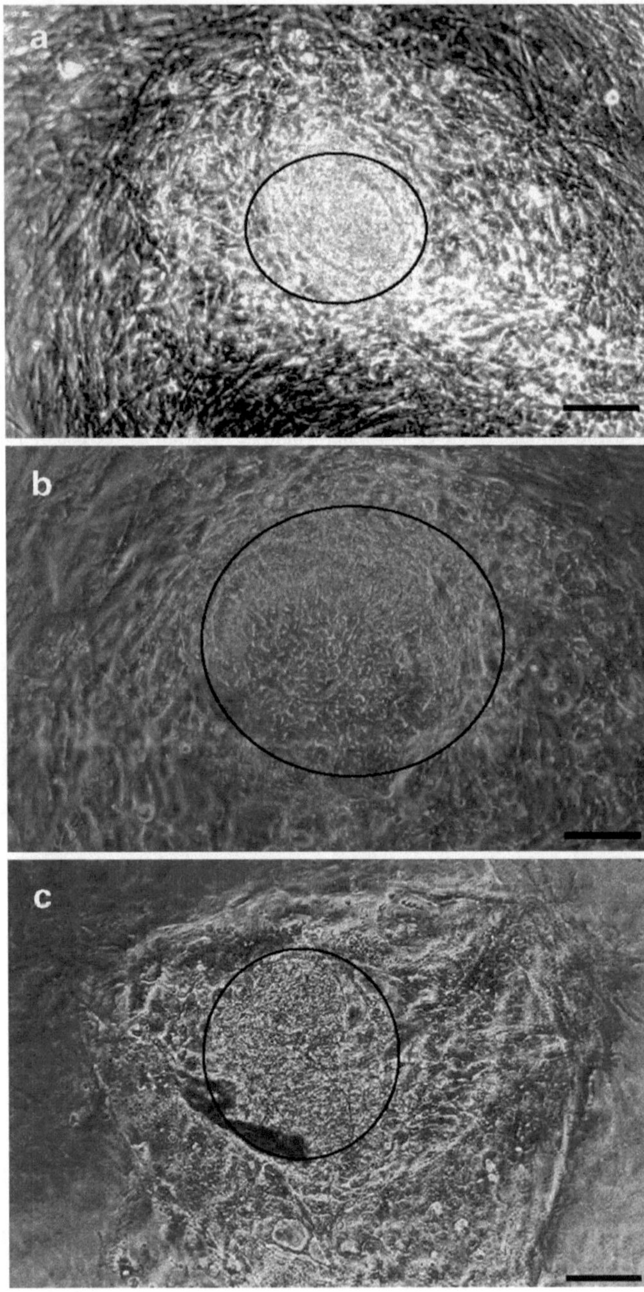

Fig. 1.11 Whole embryo approach for derivation of ESC lines—clear ICM outgrowth. (**a–c**) Three examples of plated embryos in which the ICM outgrowth can be clearly noted (marked with *circle*). Bar 90 μM

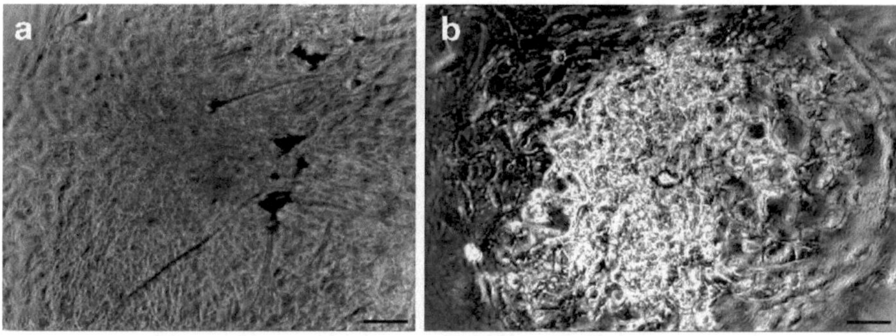

Fig. 1.12 Whole embryo approach for derivation of ESC lines on MEF—differentiation (**a**, **b**). Two examples of plated embryos with ICM differentiation, the derivation failed. Bar 70 μM and 60 μM for (**a**) and (**b**), respectively

1.2 Materials for ESC Line Derivation

1.2.1 Tyrode's acid (Sigma, acidic, C.N. T-1788).
1.2.2 Antibodies: antihuman whole antiserum (Sigma, H-8765), recommended dilution 1:30 in Dulbecco's modified Eagle's medium (DMEM).
1.2.3 Complement proteins: Guinea pig complement diluted 1:10 in DMEM or the solvent provided by the supplier (Gibco BRL C.N. 10723-013).
1.2.4 hESC—serum-based medium: 80% DMEM\F12 (DMEM, Invitrogen Corporation C.N. 10829018), 20% fetal bovine serum (FBS) (HyClone), 1% nonessential amino acid, 1 mM L-glutamine, and 0.1 mM β-mercaptoethanol.
1.2.5 hESCs—serum-free medium: hESCs can be cultured with MEFs using the following serum-free medium: 85% DMEM\F12, 15% SR (Invitrogen Corporation knockout serum replacement C.N. 10828028), 1% nonessential amino acid, 1 mM L-glutamine, 0.1 mM β-mercaptoethanol, and 4 ng/ml basic fibroblast growth factor (bFGF).

1.3 Methods for hESC Isolation

Surplus embryo should be cultured to the blastocyst stage (day 5–6 postfertilization) by a trained embryologist.

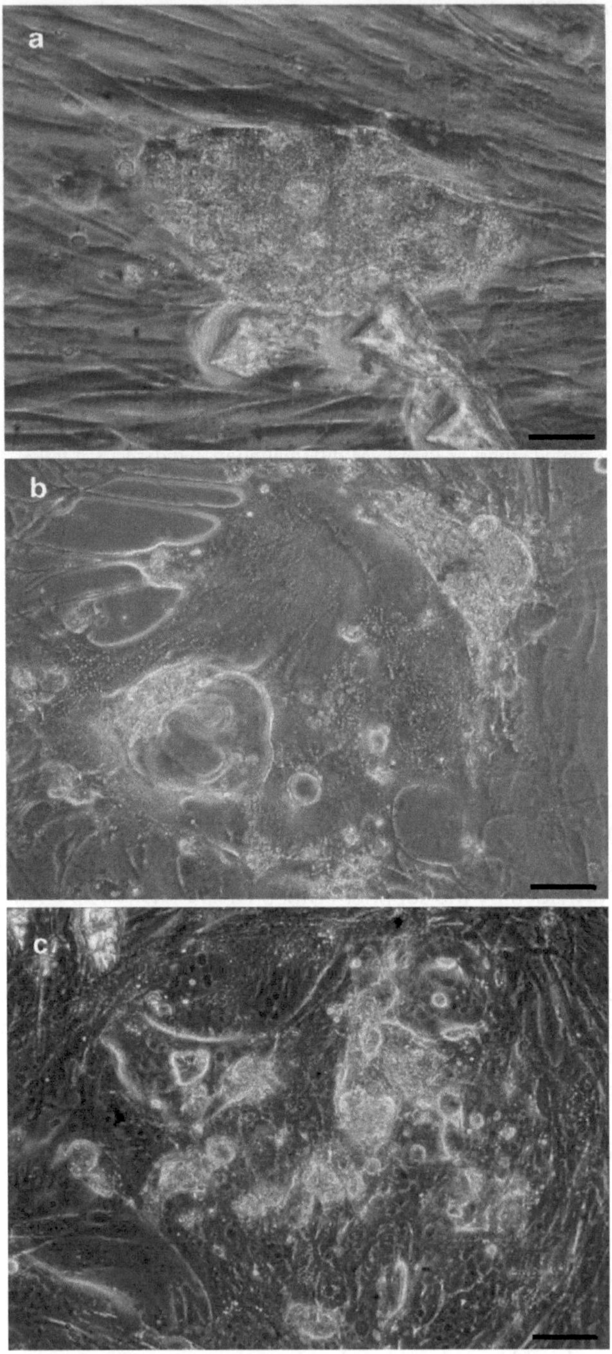

Fig. 1.13 Whole embryo approach for derivation of ESC lines on human foreskin fibroblasts—differentiation. (**a**–**c**) Failed derivation. Three examples of differentiated ICM from embryos plated on human foreskin fibroblasts. (**a**) The plated embryo resembles a colony. Since a whole embryo was plated and there is no difference between trophoblast and ICM cells, the whole colony will likely differentiate in a few days. (**b**, **c**) Clear differentiation of the plated embryo cells. Bar 60 μM and 80 μM for (**a**) and (**b**, **c**), respectively

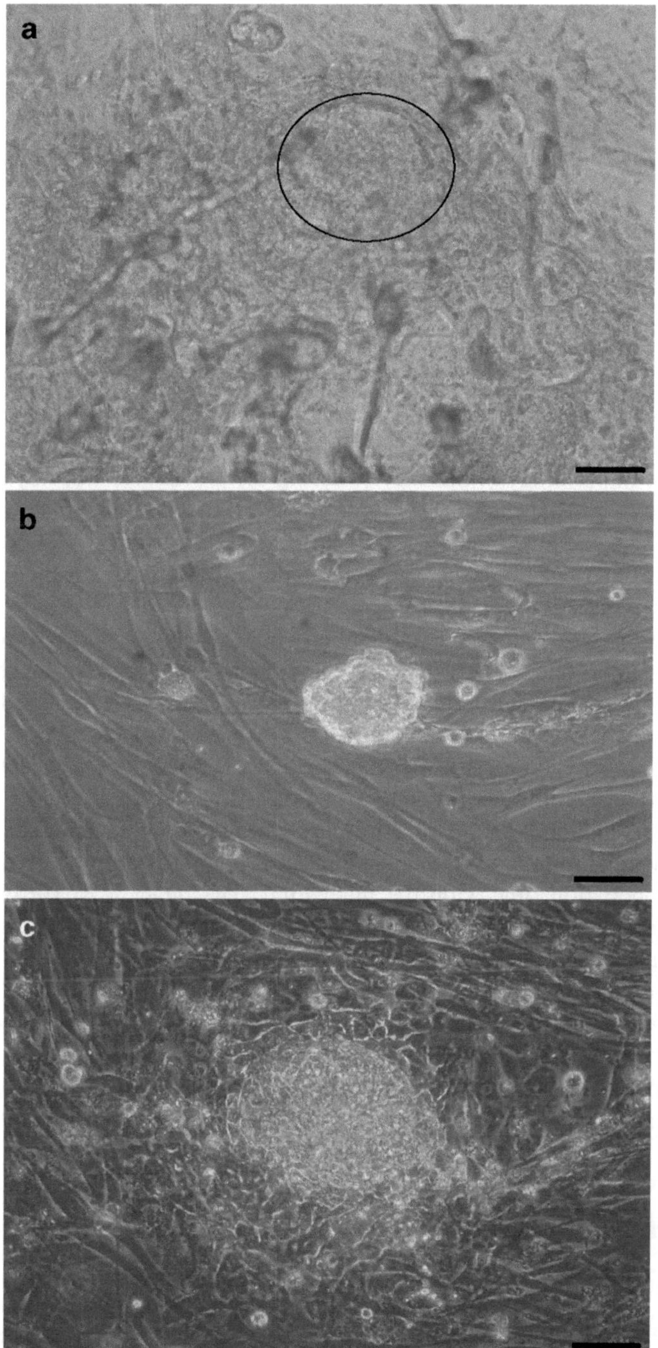

Fig. 1.14 Derivation of a GMP-grade ESC line (CL1). Implementing the whole embryo approach, an ESC line was derived in a clean room, using human foreskin fibroblasts as a feeder layer and NutriStem™ medium (animal and serum free). (**a**) Plated embryo in which ICM outgrowth can be distinguished (*circle*). (**b**) ICM after removal from the growing embryo. (**c**) The resultant ESC colony 3 days post-plating of the clean ICM. Bar 60 μM, 50 μM, and 70 μM for (**a**), (**b**), and (**c**), respectively

1.3.1 hESC Isolation by Immunosurgery

1. Prepare the following in advance: a 4-well plate covered with feeder layer containing 0.5 ml ESC medium per well for culturing (see 1.2.4 or 1.2.5); 58-mm plates with six drops of 25 μl Tyrode's acid; and three 4-well plates with 0.5 ml ESC culture medium for washing (see 1.2.4 or 1.2.5): one from the Tyrode's acid, one from the antibody, and one from the complement. The plates should be preincubated to 37°C in a culture incubator (about 10 min).
2. To remove the ZP layer, incubate the embryo for 30–60 s in a drop of previously heated Tyrode's acid (see 1.2.1). Monitor the procedure under a dissecting microscope (recommended to set a hot plate to 37°C). When the ZP starts to dissolve, which should happen within 60 s, quickly remove the embryo. Figure 1.2 illustrates the morphology of blastocysts during the procedure. Wash by transferring the embryo from well to well three times in the washing plate prepared in advance. An example of an exposed blastocyst is illustrated in Fig. 1.2b, c. Since antibodies can penetrate through the ZP, the ZP can be mechanically removed just before plating the ICM.
3. Incubate the bare embryo in antihuman whole serum antibodies (see 1.2.2) for 30 min. Figure 1.3a, b illustrates embryo morphology during incubation with antibodies. Immediately wash the embryo three times in fresh ESC medium using the washing plate prepared in advance (see 1.2.4 or 1.2.5). Precision in incubation time is not critical at this stage; variations are not expected to harm the embryo or reduce success rates.
4. Incubate the embryo for up to 20 min in guinea pig complement (see 1.2.3). It is recommended to monitor the procedure; if trophoblasts are lysed before the end of the incubation time, stop the incubation. The intact ICM surrounded with lysed trophoblasts is illustrated in Fig. 1.3c, d. Do not exceed the incubation time, long incubation can harm the ICM cells.
5. Wash the intact ICM three times in fresh ESC medium (see 1.2.4 or 1.2.5) in the washing plate prepared in advance, using a pulled pasture pipette, to remove the lysed trophoblasts.
6. Plate the intact ICM on a fresh feeder-covered culture dish (Figs. 1.3e, f, and 1.4) in ESC medium (see 1.2.4 or 1.2.5).

1.3.2 Mechanical Removal of Trophectoderm

1. Prepare the following in advance: a 4-well plate covered with feeder layer with 0.5 ml ESC medium per well for culturing (see 1.2.4 or 1.2.5), a 58-mm plate with six drops of 25 μl Tyrode's acid, and a 4-well plate with 0.5 ml ESC culture medium for washing (see 1.2.4 or 1.2.5). The plates should be preincubated to 37°C in a culture incubator for about 10 min.
2. Expose the embryo by removing the ZP as described in Sect. 1.3.1, including three washes in the 4-well plate prepared in advance. The embryo will not attach to the feeder layer if the ZP remains in place.

3. Transfer the embryo to a well in a 4-well plate covered with feeder cells. If the ICM is clearly visible, remove as much trophoblast as possible, using either 25–27 gauge syringe needles or pulled pasture pipette under a dissecting microscope. If the ICM is unrecognizable, plate the embryo as a whole (see Sect. 1.3.3). Leave the clean ICM in the same well for expansion.

1.3.3 Whole Embryo Approach for ESC Line Derivation

1. Prepare the following in advance: a 4-well plate covered with feeder layer with 0.5 ml ESC medium per well for culturing (see 1.2.4 or 1.2.5), a 58-mm plate with six drops of 25 μl Tyrode's acid, and a 4-well plate with 0.5 ml ESC culture medium for washing (see 1.2.4 or 1.2.5). The plates should be preincubated to 37°C in a culture incubator (about 10 min).
2. Expose the embryo from ZP as described in Sect. 1.3.1, including three washes in the 4-well plate prepared in advance. The embryo will not attach to the feeder layer if the ZP remains in place. An example of an embryo with ZP residue is depicted in Fig. 1.2d.
3. Transfer the embryo to a well in a 4-well plate covered with feeder cells. The embryo should attach to the feeder cells after no longer than 24 h. Figures 1.8–1.10 demonstrate different morphologies of embryos plated whole.
4. After 5–10 days, distinct ICM outgrowth should appear. Selectively cut the ICM under a dissecting microscope and transfer it to a new plate covered with a feeder layer. Examples of the morphology of ICM outgrowth are illustrated in Fig. 1.11.
5. Expand the cells. It is recommended that for the first 2–5 passages the colonies will be split mechanically, as described for ICM colonies in Sect. 1.3.3 and References. The morphology of the resulting colonies is demonstrated in Figs. 1.5–1.7.

References

Amit M, Itskovitz-Eldor J (2002) Derivation and spontaneous differentiation of human embryonic stem cells. J Anat 200:225–232

Bongso A, Fong CY, Ng SC, Ratnam S (1994) Isolation and culture of inner cell mass cells from human blastocysts. Hum Reprod 9:2110–2117

Cowan CA, Klimanskaya I, McMahon J, Atienza J, Witmyer J, Zucker JP, Wang S, Morton CC, McMahon AP, Powers D, Melton DA (2004) Derivation of embryonic stem-cell lines from human blastocysts. N Engl J Med 350:1353–1356

Evans MJ, Kaufman MH (1981) Establishment in culture of pluripotential cells from mouse embryos. Nature 292:154–156

Klimanskaya I, Chung Y, Meisner L, Johnson J, West MD, Lanza R (2005) Human embryonic stem cells derived without feeder cells. Lancet 365:1636–1641

Lanzendorf SE, Boyd CA, Wright DL, Muasher S, Oehninger S, Hodgen GD (2001) Use of human gametes obtained from anonymous donors for the production of human embryonic stem cell lines. Fertil Steril 76:132–137

Lerou PH, Yabuuchi A, Huo H, Miller JD, Boyer LF, Schlaeger TM, Daley GQ (2008) Derivation and maintenance of human embryonic stem cells from poor-quality *in vitro* fertilization embryos. Nat Protoc 3:923–933

Ludwig TE, Levenstein ME, Jones JM, Berggren WT, Mitchen ER, Frane JL, Crandall LJ, Daigh CA, Conard KR, Piekarczyk MS, Llanas RA, Thomson JA (2006) Derivation of human embryonic stem cells in defined conditions. Nat Biotechnol 24:185–187

Martin GR (1981) Isolation of a pluripotent cell line from early mouse embryos cultured in medium conditioned by teratocarcinoma stem cells. Proc Natl Acad Sci USA 78:7634–7638

Mateizel I, De Temmerman N, Ullmann U, Cauffman G, Sermon K, Van de Velde H, De Rycke M, Degreef E, Devroey P, Liebaers I, Van Steirteghem A (2006) Derivation of human embryonic stem cell lines from embryos obtained after IVF and after PGD for monogenic disorders. Hum Reprod 21:503–511

Mitalipova M, Calhoun J, Shin S, Wininger D, Schulz T, Noggle S, Venable A, Lyons I, Robins A, Stice S (2003) Human embryonic stem cell lines derived from discarded embryos. Stem Cells 21:521–526

Reubinoff BE, Pera MF, Fong C, Trounson A, Bongso A (2000) Embryonic stem cell lines from human blastocysts: somatic differentiation *in vitro*. Nat Biotechnol 18:399–404

Solter D, Knowles BB (1975) Immunosurgery of mouse blastocyst. Proc Natl Acad Sci USA 72:5099–5102

Steptoe PC, Edwards RG (1978) Birth after the reimplantation of a human embryo. Lancet 2:366

Thomson JA, Itskovitz-Eldor J, Shapiro SS, Waknitz MA, Swiergiel JJ, Marshall VS, Jones JM (1998) Embryonic stem cell lines derived from human blastocysts. Science 282:1145–7; [erratum in Science 1998;282:1827]

Verlinsky Y, Strelchenko N, Kukharenko V, Rechitsky S, Verlinsky O, Galat V, Kuliev A (2005) Human embryonic stem cell lines with genetic disorders. Reprod Biomed Online 10:105–110

Chapter 2
Morphology of Human Embryonic and Induced Pluripotent Stem Cell Colonies Cultured with Feeders

Abstract To prolong the stage of undifferentiation, human embryonic stem cells (hESCs) and induced pluripotent stem cells (iPSCs) have traditionally been isolated and cultured using feeder layers, such as mouse embryonic fibroblasts (MEFs) or foreskin fibroblasts, with medium supplemented by fetal bovine serum (FBS). For research purposes, these conditions are preferable and are often referred to as the gold standard. This chapter describes the colony morphology of undifferentiated hESCs and iPSCs cultured with MEFs or human foreskin fibroblasts.

2.1 Introduction

Traditionally, human ESCs (hESCs) are co-cultured with inactivated mouse embryonic fibroblasts (MEFs) as supporting layers, and the medium is supplemented with a high percentage of fetal bovine or calf serum (FBS) (Thomson et al. 1998). The feeder layer serves a dual role of supporting hESC expansion and preventing spontaneous differentiation. However, such conditions are not appropriate for clinical and industrial purposes due to variations between batches of FBS and MEFs, and to the risk of exposure of the cells to animal pathogens. Modifications that prevent xeno-contamination in the culture system include the use of a defined medium supplemented with serum replacement, without animal products. The MEFs should be replaced by human feeder cells or matrix.

Accumulating data demonstrate that induced pluripotent stem cells (iPSCs) can be cultured in similar conditions as those for hESCs (Takahashi and Yamanaka 2006; Takahashi et al. 2007); therefore, improvements in hESC culture conditions will also apply to those of iPSCs.

Intensive efforts have been invested during the last decade in the search for alternative feeder cells for hESCs. The result is the identification of a number of cell line types that support the culture of undifferentiated hESCs, including human fetal-derived

M. Amit and J. Itskovitz-Eldor, *Atlas of Human Pluripotent Stem Cells: Derivation and Culturing*, Stem Cell Biology and Regenerative Medicine, DOI 10.1007/978-1-61779-548-0_2, © Springer Science+Business Media, LLC 2012

fibroblasts (Richards et al. 2002), foreskin fibroblasts (Amit et al. 2003; Hovatta et al. 2003), human placenta fibroblasts (Simón et al. 2005; Genbacev et al; 2005), adult human fibroblasts (Tecirlioglu et al. 2010), and adult marrow cells (Cheng et al. 2003). All these cell lines have been demonstrated to support the prolonged culture of hESCs as undifferentiated, while maintaining all hESC features. We found human foreskin fibroblasts (HFF) to equally support the undifferentiated culture of iPSCs (Amit et al. unpublished data).

Human fetal-derived fibroblasts, placenta fibroblasts, adult human fibroblasts, and foreskin fibroblasts have also been found to support the isolation of new hESC lines under animal-free or serum-free conditions (Richards et al. 2002; Hovatta et al. 2003; Simón et al. 2005; Genbacev et al. 2005; Inzunza et al. 2005; Tecirlioglu et al. 2010). Of these, foreskin fibroblast cells are the most common, accounting for 51 of 57 (89%) of the lines reported during recent years (Richards et al. 2002; Ström et al. 2010; Aguilar-Gallardo et al. 2010; Tecirlioglu et al. 2010; Ilic et al, 2009; Valbuena et al. 2006; Ellerström et al. 2006; Genbacev et al. 2005; Simón et al. 2005; Crook et al. 2007). Of these 57 lines, the 7 reported to be clinical-grade lines were all isolated and cultured using foreskin fibroblasts as feeders (Crook et al. 2007; Ellerström et al. 2006). Thus, these cells are not only the most frequently used human feeders, but also those most ensuring a xeno-free culture system.

Though these culture systems promote animal-free conditions for culturing hESCs, they are not well defined, due to variations between batches of feeder layer cells and to the fact that some still use human serum for the feeder cell culture. An additional disadvantage to the use of human feeders is the need to culture the feeder lines, which limits large-scale culturing of hESCs. Therefore, the ideal culture method seems to be a combination of an animal-free matrix and both serum-free and animal-free medium.

2.2 Materials

2.2.1 For Mouse Embryonic Fibroblasts (MEFs) and Foreskin Fibroblasts (HFFs)

2.2.1.1 0.1% Gelatin

0.1% Gelatin (type A, from porcine, Sigma G-1890). All culture dishes should be covered with 0.1% gelatin at least 1 h before MEF plating.

2.2.1.2 Culture Medium MEFs

90% Dulbecco's modified Eagle's medium (DMEM) and 10% fetal bovine serum (FBS). During the first passage post-derivation, penicillin–streptomycin should be added (Sigma P-3539, final concentration of penicillin 10,000 u/ml and streptomycin 10 mg/ml).

2.2.1.3 Culture Medium HFFs

90% DMEM, 10% FBS, and 2 mM L-glutamine. During the first passage post-derivation, penicillin–streptomycin should be added (Sigma P-3539, final concentration of penicillin 1,000 u/ml and streptomycin 1 mg/ml).

2.2.1.4 Feeder Freezing Medium

60% DMEM, 20% dimethyl sulfoxide (DMSO), and 20% FBS.

2.2.1.5 Feeder Splitting

Trypsin/EDTA (Invitrogen Corporation, type IV C.N 17104019).

2.2.1.6 Mitomycin C

8 μg/ml mitomycin C (Sigma, M-4287) diluted in DMEM.

2.2.1.7 Washing

Phosphate-buffered saline (PBS) with Ca^{++} and Mg^{++}.

2.2.2 *For hPSC Maintenance*

2.2.2.1 hPSC: Serum-Based Medium

80% DMEM, 20% defined FBS (HyClone), 1% nonessential amino acid, 2 mM L-glutamine, and 0.1 mM β-mercaptoethanol.

2.2.2.2 hPSC: Serum-Free Medium

85% DMEM/F12, 15% knockout (KO) serum replacement (SR, Invitrogen Corporation, C.N. 10828028), 1% nonessential amino acid, 2 mM L-glutamine, 0.1 mM β-mercaptoethanol, and 4 ng/ml basic fibroblast growth factor (bFGF). For iPSCs, it is recommended to increase the bFGF concentration to 10 ng/ml.

2.2.2.3 Splitting Medium

The splitting medium consists of 1 mg/ml collagenase type IV (Wordington, type IV C.N 4189, activity of 220–320 u/mg) in DMEM.

2.2.2.4 Freezing Medium

40% DMEM, 20% DMSO, 20% FBS, and 20% SR.

2.3 Methods

2.3.1 Feeder Culture Methods

2.3.1.1 Derivation of MEFs from Pregnant Mice

1. Use of pregnant Imprinting Control Region (ICR) mice (or CD1) on the 13th day of conception is recommended.
2. Sacrifice 1 female mouse by a method approved by the ethics committee of your institution.
3. Wash the abdomen with 70% ethanol and dissect the abdominal cavity to expose the uterine horns.
4. Set the uterine horns in 10-cm^2 petri dishes and wash three times with 10 ml of PBS. A uterine horn is depicted in Fig. 2.1a.
5. Using two pairs of watchmakers' forceps (Dumont 5, Fine Scientific Tolls), open each uterine wall and release each embryo.
6. Wash retrieved embryos three times with 10 ml PBS (see see Sect. 2.2.1.7). Embryos released from the embryonic sac are illustrated in Fig. 2.1b.
7. Use the same tools to dissect each embryo from the placenta and membranes, and discard soft tissues as much as possible.
8. Transfer clean embryos into new petri dishes and mince thoroughly using sharp Iris scissors. An example of sufficiently minced embryos is depicted in Fig. 2.1c.
9. Add 6 ml of trypsin/EDTA (see Sect. 2.2.1.5) and incubate for at least 20 min.

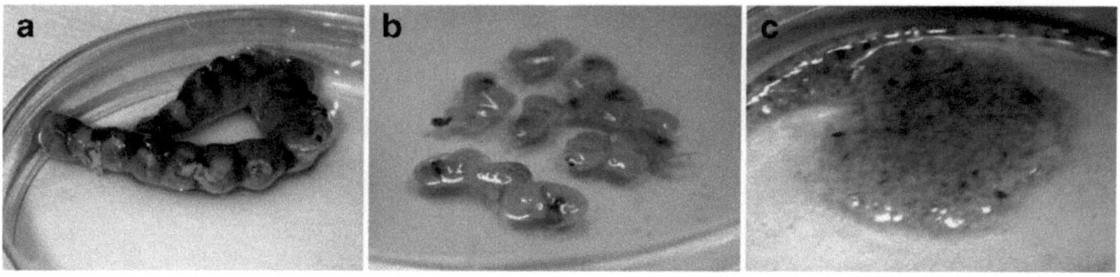

Fig. 2.1 Mouse embryonic fibroblast (MEF) derivation. (**a**) Uterine horn of ICR mice on the 13th day of conception. (**b**) Embryos released from embryonic sac. (**c**) The same culture after mincing the embryos with sharp Iris scissors to the correct sizes

10. Neutralize trypsin using at least 6 ml of MEF culture medium (see Sect. 2.2.1.2).
11. Transfer the cells into conical tubes.
12. Divide evenly into T75 culture flasks. We recommend a ratio of three embryos per flask.
13. Add 20 ml of MEF culture medium to each flask (see Sect. 2.2.1.2).
14. Grow the MEFs up to 3 days or until the culture is confluent. Change the medium at least once during culturing (do not aspirate any floating clumps). Morphology of derived MEFs during the first days of culture post-derivation is illustrated in Fig. 2.2.
15. Freeze the resulting MEF (see Sect. 2.3.1.4).

2.3.1.2 HFF Derivation

1. A newborn human foreskin should be placed in PBS supplemented with penicillin–streptomycin immediately following the circumcision. The foreskin should be maintained at 4–8°C until derivation of fibroblasts, no later than 48 h post circumcision.
2. Unfold the foreskin and wash three times with PBS (see Sect. 2.2.1.7).
3. Cut into small pieces using sharp Iris scissors (about eight pieces per foreskin).
4. Transfer clean pieces into a new petri dish and mince thoroughly using sharp Iris scissors.
5. Add 6 ml of trypsin/EDTA (see Sect. 2.2.1.5) and incubate for at least 30 min.
6. Neutralize the trypsin using at least 6 ml of HFF culture medium (see Sect. 2.2.1.3). Transfer the HFF into conical tubes. Use HFF culture medium to wash the plate.
7. Divide evenly into T25 culture flasks at a recommended ratio of two pieces per flask.
8. Add 6 ml of HFF culture medium (see Sect. 2.2.1.3).
9. Grow the HFF until the culture is confluent. Change medium as needed, every 5 days if not split. HFF morphology is demonstrated in Fig. 2.4.

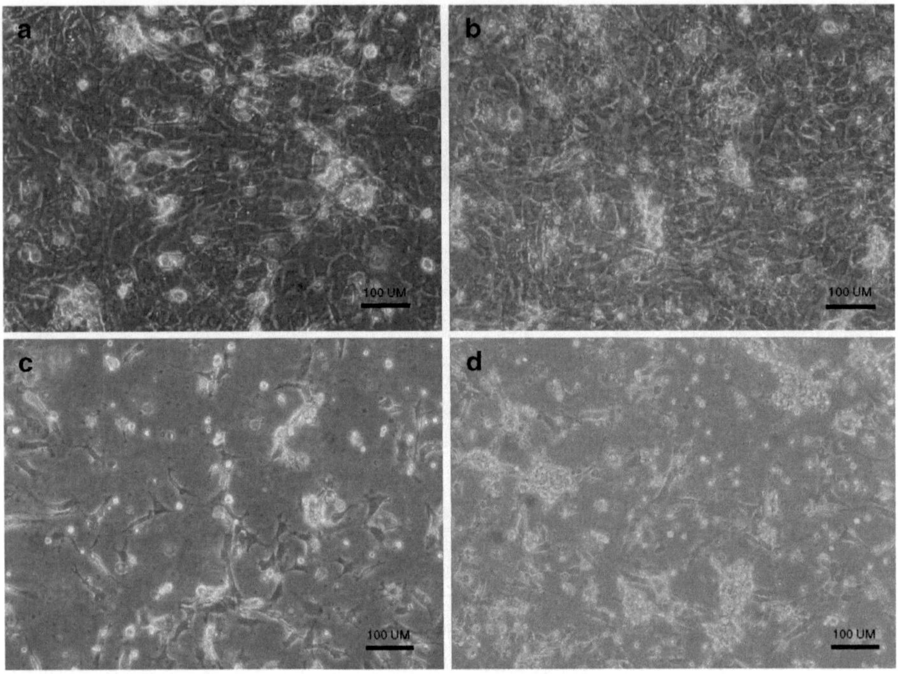

Fig. 2.2 MEF primary culture 2 days post-derivation. (**a, b**) Examples of MEF cultures with expected concentration and fibroblast morphology. (**c, d**) Examples of MEF cultures with poor recovery; the culture confluence is less than 40%. Bar 100 μM

2.3.1.3 Feeder Splitting

1. Aspirate the culture medium and wash the flask once with 5 ml PBS (for T75 flask, see Sect. 2.2.1.7).
2. Add 2 ml of trypsin/EDTA (see Sect. 2.2.1.5) and cover the entire culture flask surface.
3. Incubate for 6 min.
4. Tap the side of the flask to loosen the cells. Add 4 ml of culture medium (see Sects. 2.2.1.2 or 2.2.1.3) to neutralize the trypsin.
5. Transfer the cell suspension into a conical tube and centrifuge for 5 min at $90 \times g$.
6. Remove the suspension from the centrifuge, re-suspend in 2 ml of culture medium (see Sects. 2.2.1.2 or 2.2.1.3), and pipette to fracture the pellet.
7. Distribute the cell suspension to a desired number of culture flasks. For MEFs, we recommend a ratio of 1:5 at passage 1, 1:4 at passage 2, and 1:3 at passages 3–5; and for HFF, a ratio of 1:3.
8. Add culture medium (see Sects. 2.2.1.2 or 2.2.1.3) to reach a final volume of 10 ml.

2.3.1.4 Feeder Freezing

Once inactivated, feeder cells can be frozen at a confluence of at least 80% of the culture dish.

1. Wash flasks once with 5 ml PBS (for T75, see Sect. 2.2.1.7) and remove aggregates as much as possible.
2. Add 2 ml of trypsin/EDTA (see Sect. 2.2.1.5) and cover the entire culture flask surface.
3. Incubate for 6 min.
4. Tap the side of the flask to loosen cells. Add 4 ml of culture medium (see Sects. 2.2.1.2 or 2.2.1.3) to neutralize the trypsin.
5. Transfer the cell suspension to a conical tube. Let the remaining aggregates sink (1–2 min) and transfer the cell suspension to a clean conical tube.
6. Centrifuge for 5 min at $90 \times g$.
7. Remove the suspension, re-suspend in 2 ml culture medium (see Sects. 2.2.1.2 or 2.2.1.3), and pipette to fracture the pellet.
8. Add, drop by drop, an equivalent volume of freezing medium (see Sect. 2.2.1.4) and mix gently. Adding the freezing medium drop by drop is crucial for cell recovery.
9. Place 1 ml of the medium into 2-ml cryogenic vials. A concentration of 1–2 million cells per vial is recommended.
10. Freeze vials overnight at −80°C in a freezing box (Nalgene freezing box. C.N.5100-0001) for at least 24 h, but for no more than 1 week.
11. Transfer the vials into a liquid nitrogen container.

2.3.1.5 Feeder Thawing

1. Remove the vial from liquid nitrogen and thaw briefly in a 37°C water bath.
2. When a small pellet of frozen cell remains, clean the vial using 70% ethanol.
3. Pipette the contents of the vial once and transfer the cells into a conical tube.
4. Add, drop by drop, 2 ml of culture medium (see Sects. 2.2.1.2 or 2.2.1.3). Adding the medium drop by drop is crucial for cell recovery.
5. Centrifuge for 5 min at $90 \times g$.
6. Re-suspend the pellet in culture medium (see Sects. 2.2.1.2 or 2.2.1.3).
7. Transfer the cell suspension to culture flasks and add 10 ml of culture medium (see Sects. 2.2.1.2 or 2.2.1.3). A ratio of 1–2 million frozen cells to one T75 flask is recommended.

2.3.1.6 Preparation of Feeder-Covered Plates

1. Add 8 μg/ml of mitomycin C (see Sect. 2.2.1.6) to culture flasks and incubate for 2 h. Alternatively, feeder cells can be irradiated at 35 grays gamma irradiation.
2. Wash four times with 10 ml PBS (see Sect. 2.2.1.7).

3. Add 2 ml of trypsin/EDTA (see Sect. 2.2.1.5) and cover the entire culture flask surface (T75).
4. Incubate for 6 min.
5. Tap the side of the flask to loosen cells. Add 4 ml of culture medium (see Sects. 2.2.1.2 or 2.2.1.3) to neutralize the trypsin.
6. Transfer the cell suspension to a conical tube.
7. Centrifuge for 5 min at $90 \times g$.
8. Remove the suspension, re-suspend in 10 ml of culture medium (see Sects. 2.2.1.2 or 2.2.1.3), and pipette to fracture the pellet.
9. Count cells and re-suspend in a medium of the desired volume (see Sects. 2.2.1.2 or 2.2.1.3).
10. Transfer the cell suspension to culture dishes previously covered with gelatin (see Sect. 2.2.1.1). We recommend 4×10^5 cells per well in 6-well plates (10 cm^2) per 2 ml. Figure 2.3a, b illustrates MEF concentrations; identical concentrations should be used with HFF.
11. Let set for at least 2 h before plating hESCs. It is recommended to prepare the plates 1 day before use. Examples of MEF monolayer morphology are illustrated in Fig. 2.3c–f; and of HFF in Fig. 2.4.

2.3.2 hPSC Culture

The same culture methodology is used for MEFs and HFFs. Culture methods for hESCs and hiPSCs are also the same, other than the culture medium (see Sect. 2.2.2.2).

2.3.2.1 hPSC Splitting

The culture should be split every 4–6 days when using serum-free medium and every 5–7 days when using serum containing medium. Examples of ready-to-split colonies are depicted in Fig. 2.5.

1. Aspirate the medium from the wells that are to be split. Add splitting medium (see Sect. 2.2.2.3) to cover the wells (0.5 ml for 10 cm^2) and incubate for 20–40 min. Most colonies will float. The morphology of colonies during incubation is shown in Fig. 2.6.
2. Add 1 ml of culture medium (see Sects. 2.2.2.1 or 2.2.2.2) and gently collect the floating cells. Most feeder cells will remain behind, as exemplified in Fig. 2.7.
3. Collect the cell suspension and place into a conical tube.
4. Centrifuge for 3 min at $80 \times g$ at a recommended temperature of 4°C.
5. Aspirate the medium from fresh MEF-covered plates, re-suspend cells in medium, and plate. The size of the resulting clumps, including examples of clumps broken to incorrect sizes, is illustrated in Fig. 2.8. Too small clumps may

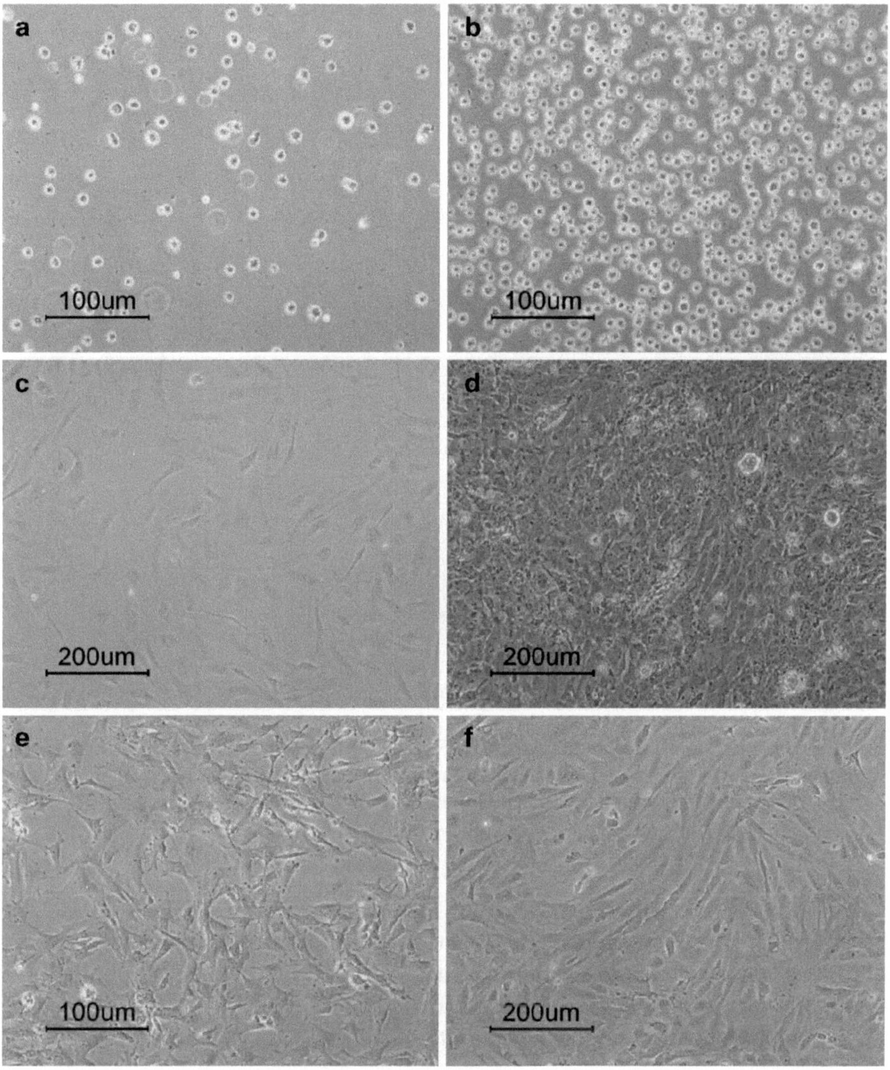

Fig. 2.3 Preparation of MEF-covered plates. (**a**) MEFs after tripsinization, demonstrating a low concentration, which may be insufficient to support pluripotent stem cell (PSC) undifferentiated culture. (**b**) MEFs after tripsinization, demonstrating a high concentration, which will probably result in a plate suitable to support PSC undifferentiated culture. (**c**) Inactivated MEF-covered plate, demonstrating low concentration, which may be insufficient to support PSC undifferentiated culture. (**d**) A plate demonstrating a very high concentration of MEF, which supports the culture of PSCs , but which may detach from the plate after a few days of growth. (**e**) Inactivated MEF-covered plate, demonstrating a low concentration, which is probably sufficient to support PSC undifferentiated culture. (**f**) A plate demonstrating the correct concentration of MEF for supporting PSC culture. (**a**, **b**, **e**) Bar 100 μM, (**c**, **d**, **f**) bar 200 μM

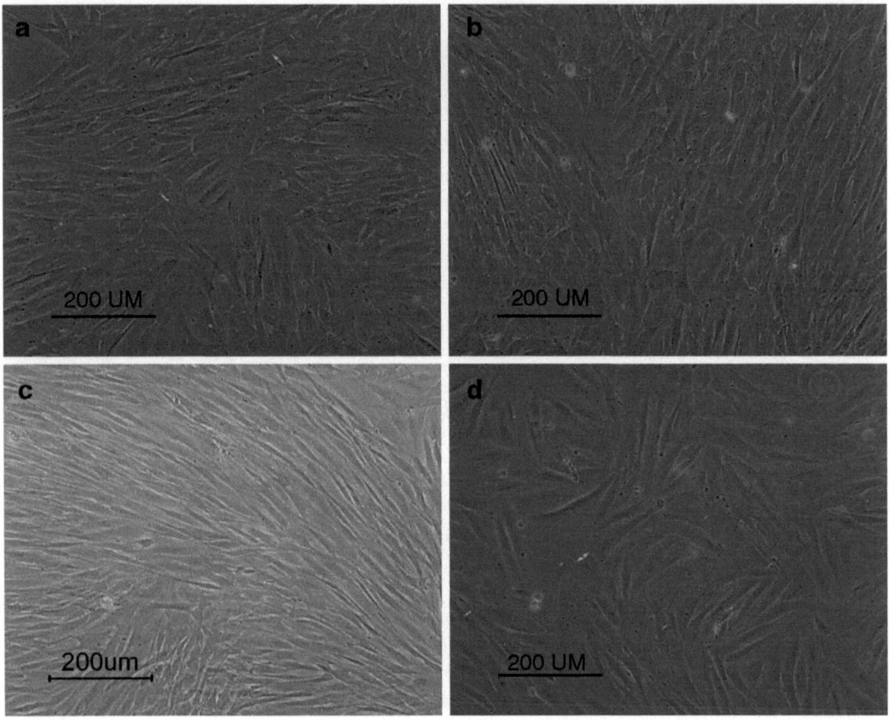

Fig. 2.4 Preparation of foreskin fibroblast (HFF)-covered plates. (**a**, **b**) Cultured HFF; note that the cells have narrow and more homogenous morphology than MEFs. (**c**, **d**) Inactivated HFF with sufficient concentration to support PSC culture. Bar 200 μM

result in decreased cell survival. If the post-splitting clump size is too large, cell attachment to the feeder cells may be harmed, resulting in increased background differentiation and greater splitting rates. Examples of colony morphology for clumps that are too large are depicted in Figs. 2.9 and 2.10c–f. A general view of colony morphology at 1 day post-splitting is illustrated in Fig. 2.10.

2.3.2.2 hPSC Freezing

1. Aspirate medium from wells to be split. Add splitting medium (see Sect. 2.2.2.3) to cover the wells (0.5 ml for 10 cm^2) and incubate for 20–40 min. Most colonies will float.
2. Add 1 ml of culture medium (see Sects. 2.2.2.1 or 2.2.2.2) and gently collect the floating cells.
3. Collect the cell suspension and place into a conical tube.
4. Centrifuge for 3 min at $80 \times g$ at a recommended temperature of 4°C.

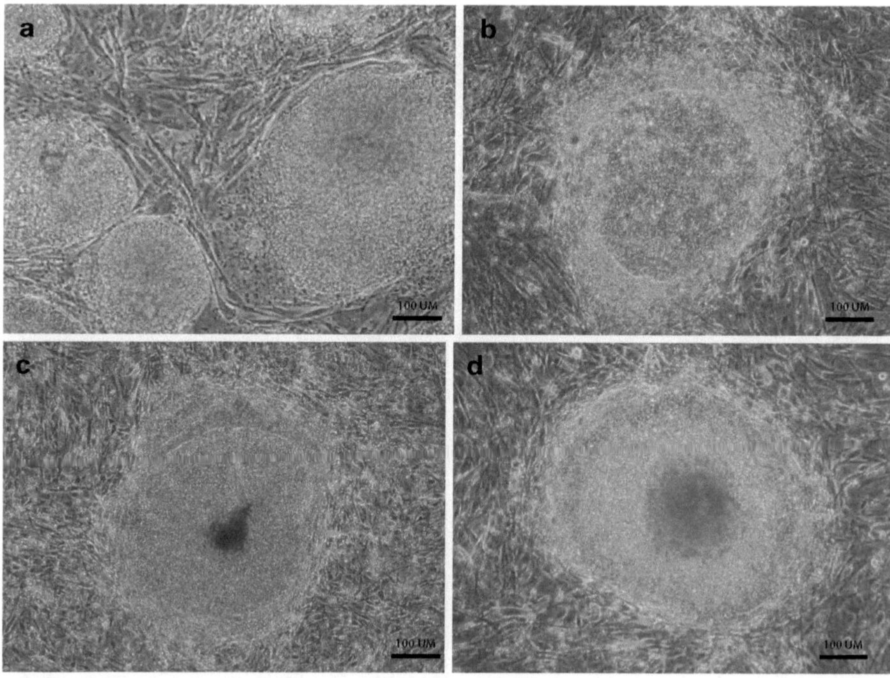

Fig. 2.5 PSCs ready for splitting. (**a**) BG01 hESCs demonstrating confluent culture, which should be split to prevent differentiation. The colony size is sufficient to survive splitting. (**b–d**) iLBWT30m hiPSC colonies (from skin biopsy, O. Brustle, Bonn University) of sufficient size for splitting; the colonies demonstrated in (**c**) and (**d**) may differentiate if splitting will be delayed. Bar 200 μM

5. Re-suspend cells in a culture medium (see Sects. 2.2.2.1 or 2.2.2.2).
6. Add, drop by drop, an equivalent volume of freezing medium (see Sect. 2.2.2.4) and mix gently. Adding the freezing medium drop by drop is crucial for cell recovery.
7. Pour 0.5 ml into a 1-ml cryogenic vial. A freezing ratio of cells covering 10 cm^2 of culture per vial is recommended.
8. Freeze overnight at −80°C in freezing boxes (Nalgene freezing box. C.N.5100-0001).
9. Transfer to liquid nitrogen on the following day.

2.3.2.3 hPSC Thawing

1. Remove the vial from the liquid nitrogen.
2. Gently swirl the vial in a 37°C water bath.
3. When a small pellet of frozen cells remains, wash the vial in 70% ethanol.

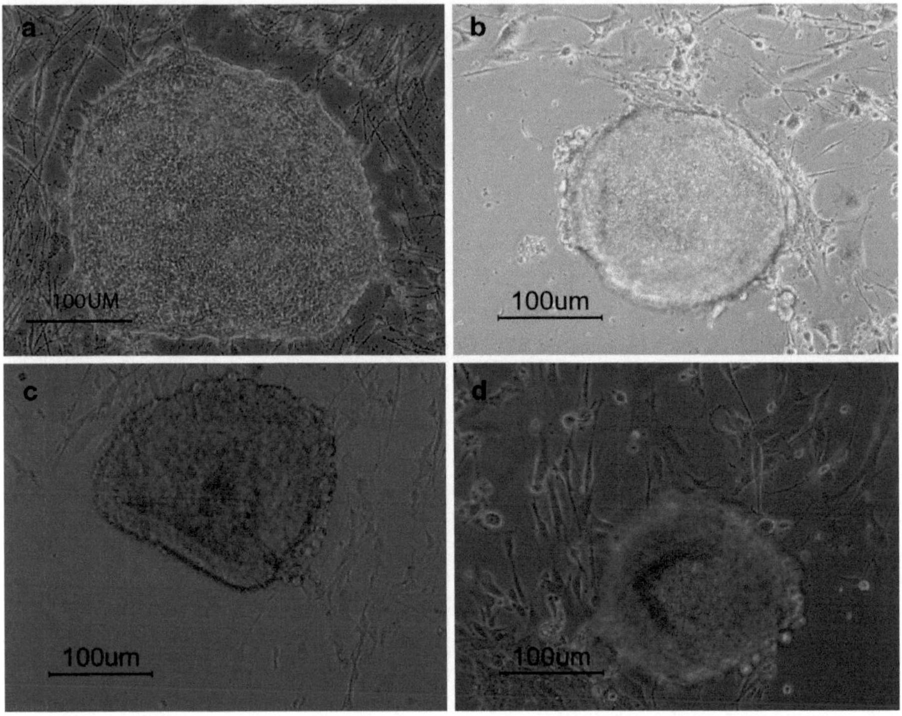

Fig. 2.6 PSCs incubated with collagenase. (**a**) Colony starting to detach from MEFs. (**b**, **c**) Colonies partly separated from MEF feeder layer; at this stage, the incubation with collagenase can be stopped, and the colonies will then easily separate from the feeder layer. (**d**) Floating colony. Note that most MEFs that remained attached themselves to the culture dish. Bar 100 μM

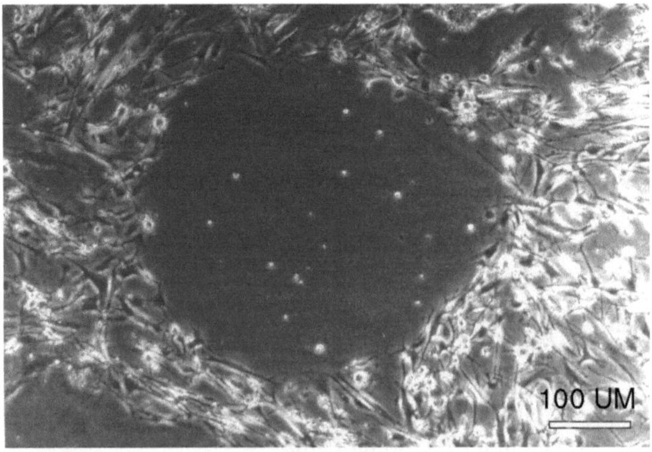

Fig. 2.7 MEFs post-collagenase splitting. MEFs that were still attached after collagenase splitting. Note the hole in the feeders created by detached PSC colonies. Bar 100 μM

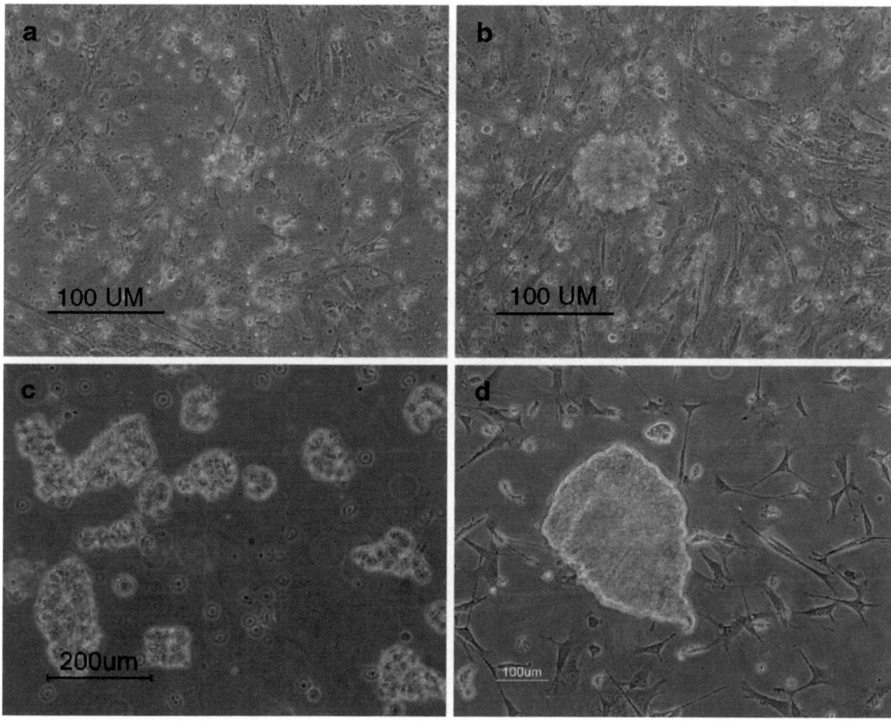

Fig. 2.8 The size of post-splitting clumps. PSC clumps resulting from collagenase treatment. (**a**) Small clumps may result in poor survival. (**b, c**) Clump size that ensures good cell recovery. (**d**) Large clumps may result in cell differentiation. Bar 100 μM, for all but (**c**), which is 200 μM

4. To mix, pipette the contents of the vial up and down once.
5. Place the contents of the vial into a conical tube and add, drop by drop, 2 ml of culture medium (see Sects. 2.2.2.1 or 2.2.2.2). Adding the culture medium drop by drop is crucial for cell recovery. Examples of the resultant clumps are illustrated in Fig. 2.11.
6. Centrifuge for 3 min at 80×g at a recommended temperature of 4°C.
7. Remove the supernatant and re-suspend the cells gently in 2 ml medium.
8. Place the cell suspension in one well of a 6-well plate, or of a 4-well plate, covered with feeder cells (see Sect. 2.3.1.5).

If the thawing procedure succeeds, small colonies should appear 1 day post-thawing (Fig. 2.12). With very good recovery, as demonstrated in Fig. 2.13, colonies will continue to grow, and will not differ from post-splitting cells. However, in some cases, when either freezing or thawing fails, the cells might not survive, as can be seen in the examples illustrated in Fig. 2.14. In some cases, the feeder clumps survive the freeze and thaw cycle, resulting in a feeder cell colony (Fig. 2.14a, c).

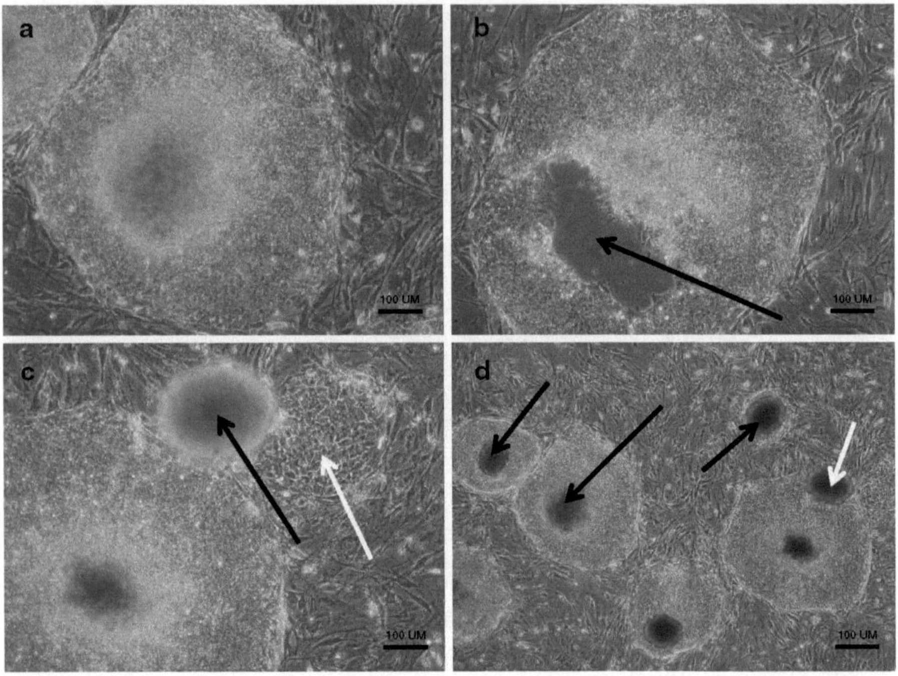

Fig. 2.9 Colonies of the I3.2 hESC line 2 days post-splitting, after cells were not broken into small enough pieces during the splitting procedure. (**a**) Since the colonies start from big clumps, they reach sufficient size for splitting 2 days later. (**b**) The colony diameter at day two is so large that a rapture was formed (marked with *arrow*). (**c**) The culture also contains colonies with small clumps (marked with *white arrow*), which may not survive the following passage. The passage should be performed when most other colonies of the culture will be suitable for splitting. Some clumps failed to attach to the feeder layer (marked with *black arrow*). (**d**) Some colonies contain big clumps at the center, which did not attach properly to the feeder layer (marked with *black arrow*). These clumps may differentiate, or the cells may become apoptotic during the passage. Some clumps settle on a different colony (marked with *white arrow*). Bar 100 μM

Occasionally, colony morphology is disorganized post-splitting and a normal-looking colony appears only a passage or two later (Fig. 2.15). As with splitting, if during the freeze and thaw cycle the cells are broken into too small clumps, they might not survive, or alternatively colonies will appear only a few days post-thawing (Fig. 2.16). After good recovery, cell concentration will resemble splitting and some clear differentiation may appear (Fig. 2.17). Even after good recovery, some cells within the colony may not survive, and dying cells may appear, usually with no harm to cell recovery (Fig. 2.18).

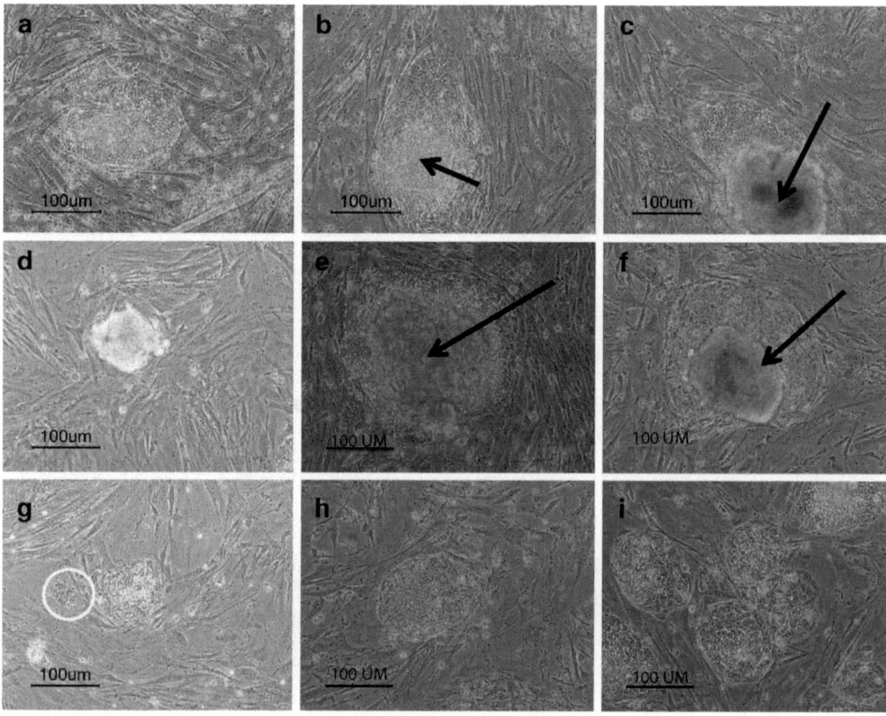

Fig. 2.10 Resultant PSC colonies 1 day post-splitting. (**a**) BG01 undifferentiated hESC colony. (**b**) BG01 undifferentiated colony with a small clump that did not attach exactly (marked with *black arrow*), and which will probably disappear after a day or two of culture with no harm. (**c**) BG01 hESC colony with a large detached clump at the center (marked with *black arrow*), which may differentiate. (**e, f**) Colony morphology is similar to that of I3 hESC line. (**d**). Small clump of BG01 hESC attached to the MEFs that have not yet demonstrated outgrowth. (**g**) Small colonies resulting from small clumps of H9.2 hESCs, one of which is hard to recognize (marked with a *circle*). (**h, i**) Additional examples of undifferentiated hESC colonies from H9.2 hESC. Bar 100 μM

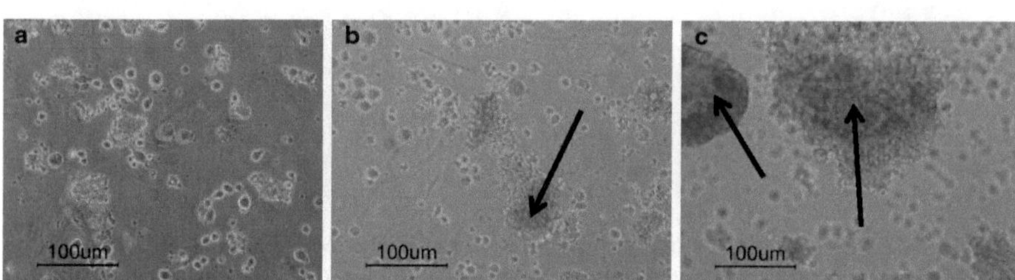

Fig. 2.11 Post-thaw clumps of I6 hESCs, before attachment to the feeder layer. (**a**) Small clump size and (**b**) good clump size (marked by *arrow*). (**c**) Large clumps with *brown* color (marked by *arrows*), may contain apoptotic cells. Bar 100 μM

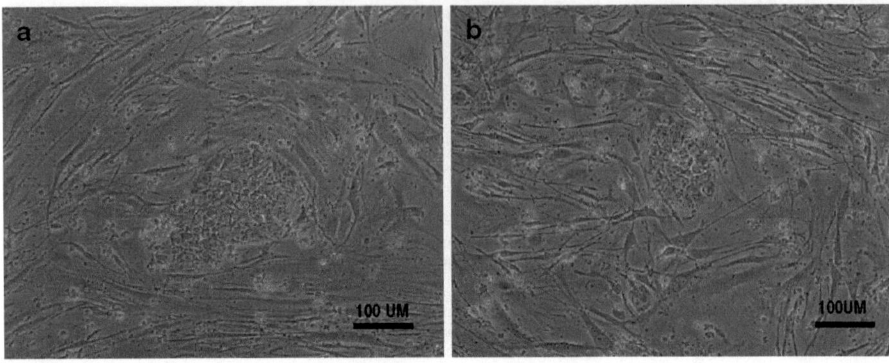

Fig. 2.12 Undifferentiated hESC colonies 1 day post-thawing. (**a**, **b**) H9.2 undifferentiated colonies surviving freeze and thaw cycle. Bar 100 μM

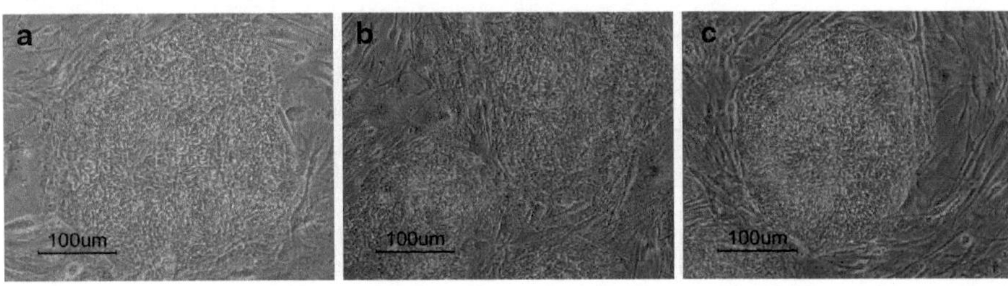

Fig. 2.13 Undifferentiated hESC colonies 2 days post-thawing. (**a–c**) I6 undifferentiated colonies surviving freeze and thaw cycle, with greater than average recovery rate. Bar 100 μM

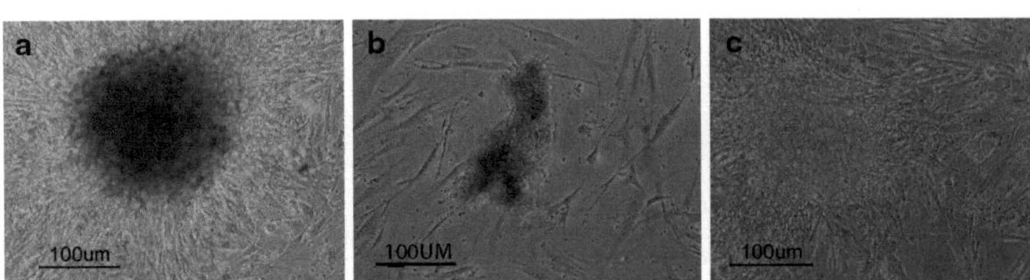

Fig. 2.14 MEF post-thawing recovery and dead cells. (**a**) A large MEF clump that survived thawing; the *brown* area probably contains apoptotic cells. (**b**) Attached clump containing dead cells; note the *brown* color. (**c**) MEF clump that survived thawing. Note the edge of the attached clump; only fibroblasts can be recognized at the outgrowth. Bar 100 μM

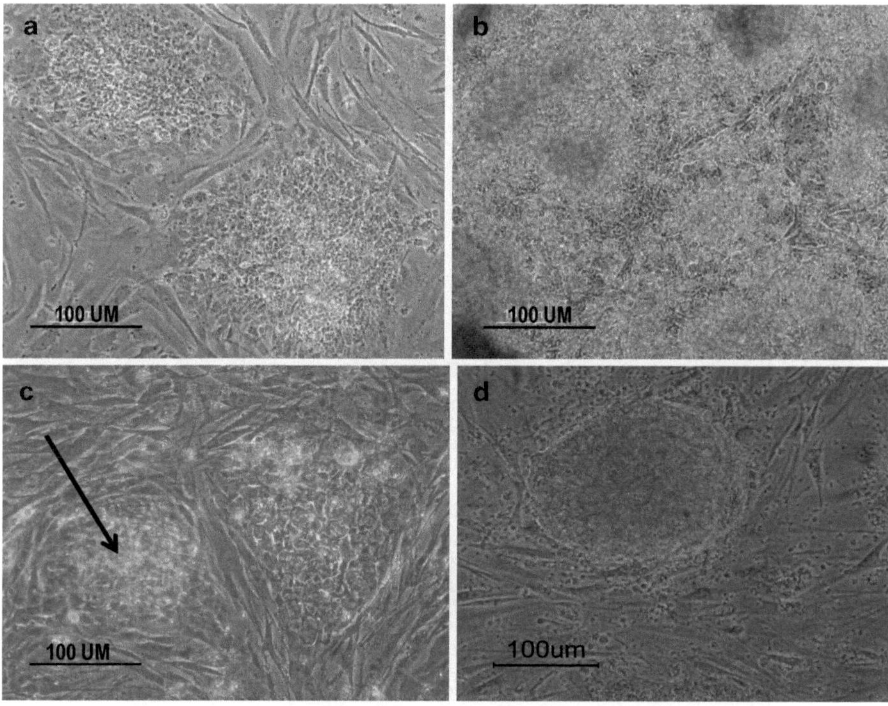

Fig. 2.15 Post-thawing recovered colonies, with disorganized morphology. (**a**) Two colonies of hESC I4, with unclear status of differentiation. (**b**) Colonies formed from hESC I4, in which all the colonies are located at the center of the dish, without clear borders and cell morphology. This can happen after thawing or splitting if the plates are not tilted before putting them in the incubator. (**c**) A hESC I6 colony with undifferentiated morphology and a second one with unclear morphology (marked with *arrow*). (**d**) A thick colony formed from hESC I6. The multiple layers make it hard to determine the cell morphology. It is not unusual for post-thawing cells to have disorganized morphology; it is, therefore, recommended to culture them for 2–3 passages before assessing and using them. Bar 100 μM

2.3.2.4 Routine Culture of hESCs

The medium (see Sects. 2.2.2.1 or 2.2.2.2) should be changed daily. If this is not possible, its quantity should be doubled. hESCs should be passaged (see Sect. 2.3.2.1) directly on fresh feeder-covered plates (see Sect. 2.3.1.6), every 4–6 days if serum-free medium is used (see Sect. 2.2.2.2) or every 5–7 days if serum containing medium is used (see Sect. 2.2.2.1).

Undifferentiated PSC colonies typically have clear borders from the feeders and contain small round cells, with spaces between them, and large nuclei with notable nucleoli. The undifferentiated morphology of PSCs cultured with MEFs or HFFs is illustrated in Figs. 2.19 and 2.20, respectively. Cell morphology within undifferentiated colonies is demonstrated in Fig. 2.21. In general, any colony with a change in

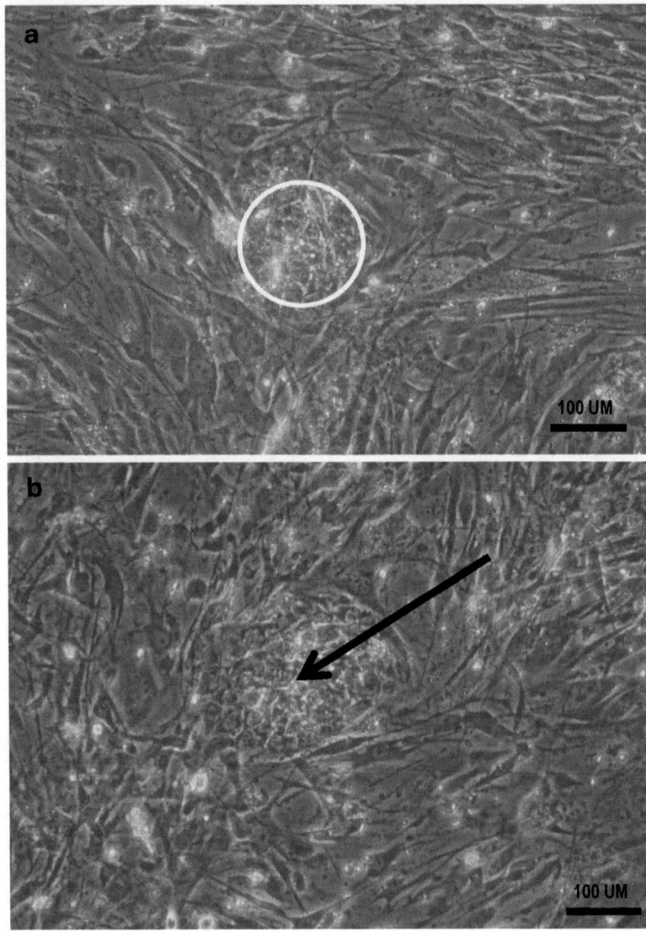

Fig. 2.16 Colonies formed from hESC I6.2 4 days post-thawing. (**a**) A tiny colony that is hardly detectable (marked with *circle*). (**b**) A clear undifferentiated colony (marked with *arrow*), relatively small for 4 days of culture. If the cells are broken into too small clumps while freezing or thawing, some clumps will not survive; however, the surviving ones will appear a few days post-thawing. It is, therefore, recommended to maintain the culture of thawed cells for 1 week to allow for the appearance of surviving colonies. Bar 100 μM

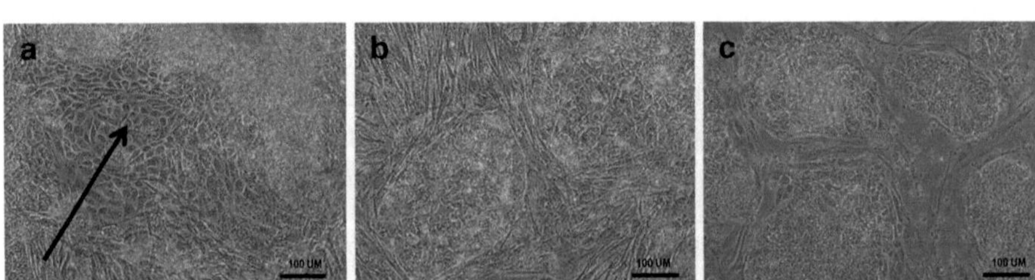

Fig. 2.17 Colonies from hESC H9 5 days post-thawing. (**a**) Cells are already differentiated (marked with *arrow*), differentiation probably occurred upon passage to freezing. (**b, c**) Undifferentiated colonies with good recovery and size. Bar 100 μM

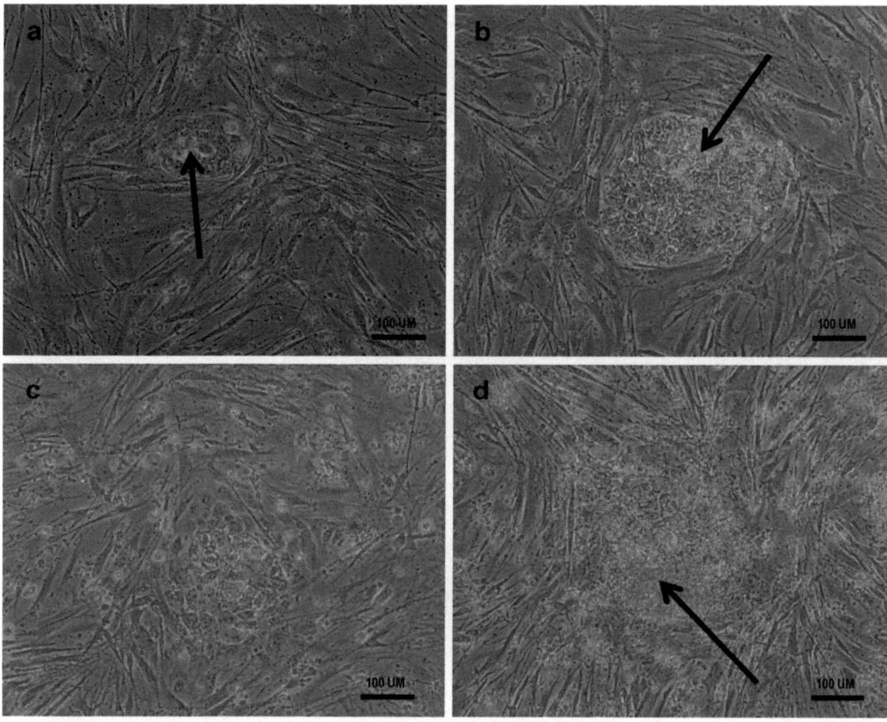

Fig. 2.18 Recovered colonies 2 days post-thawing from H9.2 hESCs, with disorganized morphology and some apoptotic cells. (**a**, **b**, **d**) Undifferentiated colonies with apoptotic cells (marked by *arrow*) detaching from the colony. In most cases, apoptotic cells will disappear after a few days, and the surviving cells will remain unchanged. (**c**) Additional examples of colonies with disorganized morphology. Bar 100 μM

its typical morphology can be considered differentiating; however in some cases, this is hard to determine without pluripotency marker testing (Fig. 2.22). The morphology of differentiated PSCs cultured with MEFs or HFF is illustrated in Figs. 2.23 and 2.24, respectively. Sometimes, a colony remains mostly undifferentiated, with differentiation starting in small areas at its edges (Fig. 2.25). Other colonies have clear differentiation morphology with distinct formation of structures (Fig. 2.26). It is recommended to scrape differentiating colonies after every 5–7 passages.

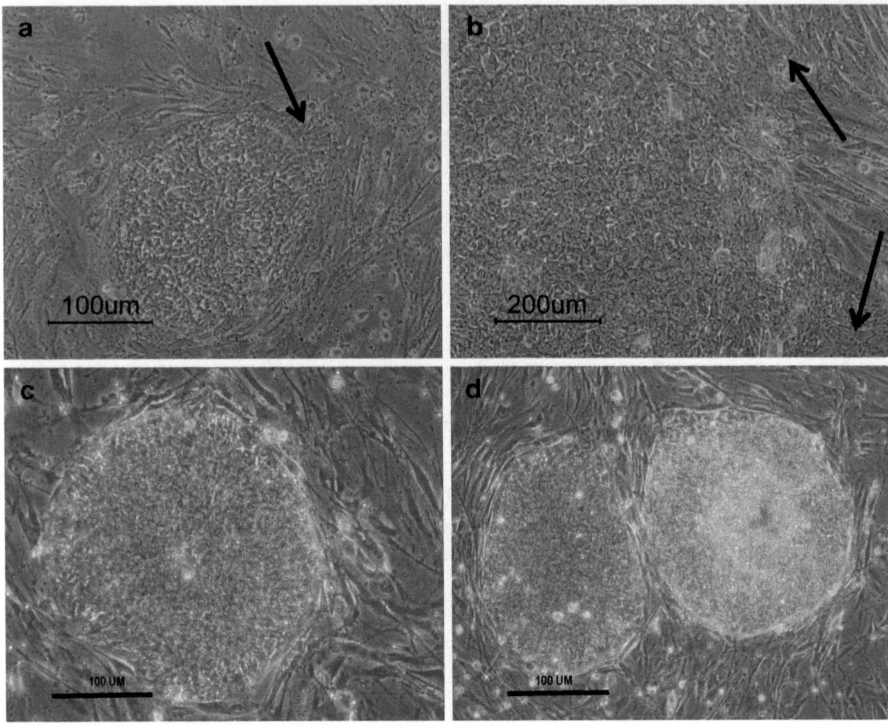

Fig. 2.19 Undifferentiated colonies. (**a**) A colony 4 days post-splitting formed by hESC BG01, in which spaces between cells can be detected. There is a small area (marked by *arrow*) where some cells slip away from the colony border with the MEFs, but these cells still have clear undifferentiated morphology. (**b**) A colony 3 days post-splitting formed from hESC H9.2, in which spaces between cells can be detected, as well as the typically large nucleus. Similar to the colony presented in (**a**). There are small areas (marked by *arrow*) where some cells slip away from the colony border, demonstrating still clear undifferentiated morphology. It seems that this large colony was formed by two nearby colonies. (**c, d**) Round undifferentiated colonies formed from hESC I3.2 2 days post-splitting, with clear borders from the MEF feeder layer. (**a, c, d**) Bar 100 μM, (**b**) bar 200 μM

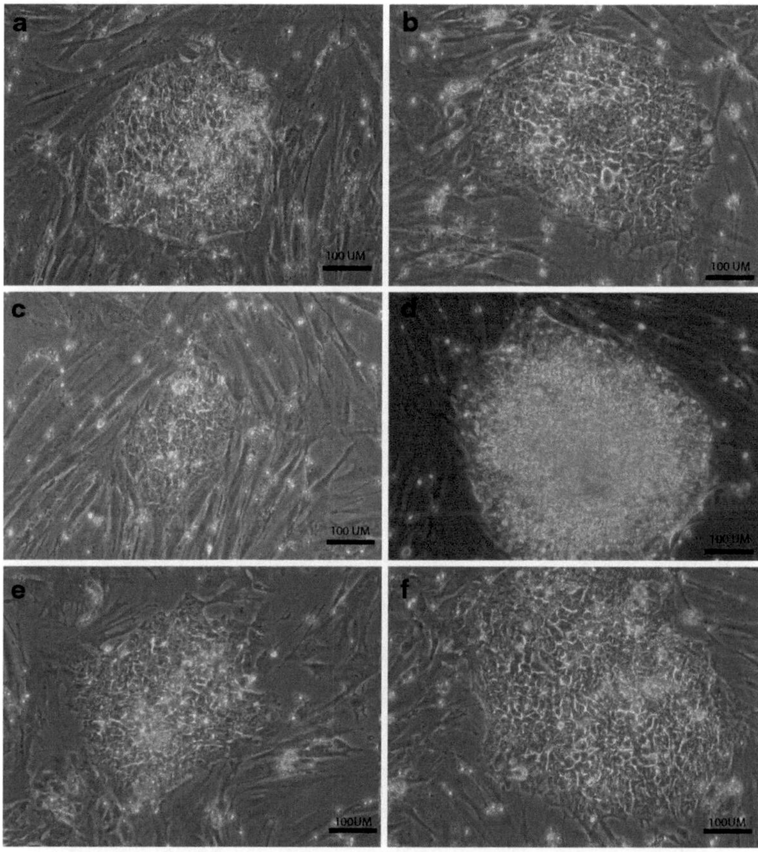

Fig. 2.20 Undifferentiated colonies formed from hESC I3.2 cultured on HFF and with serum- and animal-free medium (NutriStem™, Biological Industries Ltd). The morphology of most colonies is similar to that of cells cultured with MEFs (**a**, **c**, **d**), but some colonies (**b**, **e**, **f**) do not have clear borders, although the cells within the colony remain undifferentiated. Bar 100 μM

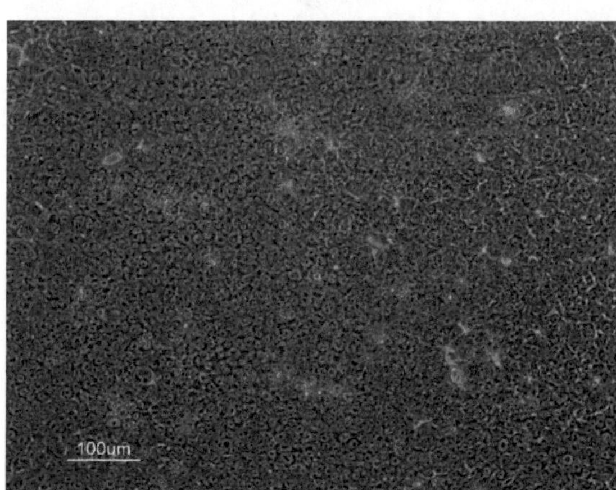

Fig. 2.21 PSC morphology. Morphology of hESC H9.2 within the colony. Small round cells with large nuclei, notable nucleoli, and spaces between cells. Bar 100 μM

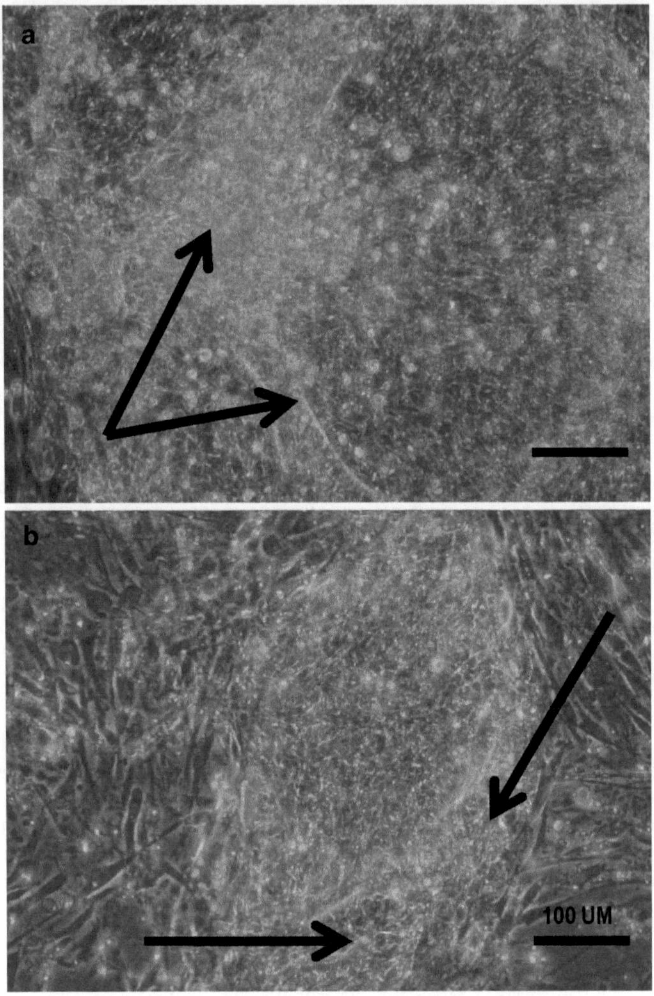

Fig. 2.22 Colonies formed from hiPSC iLBWT30m with unclear morphology. (**a**) Probably three undifferentiated colonies that combined to one. The colony borders apparently form a differentiated structure (marked by *arrow*). (**b**) Colony with cells that slip away from the colony border with unclear morphology (*arrow*). The cells within the colony have typical undifferentiated morphology. Bar 100 μM

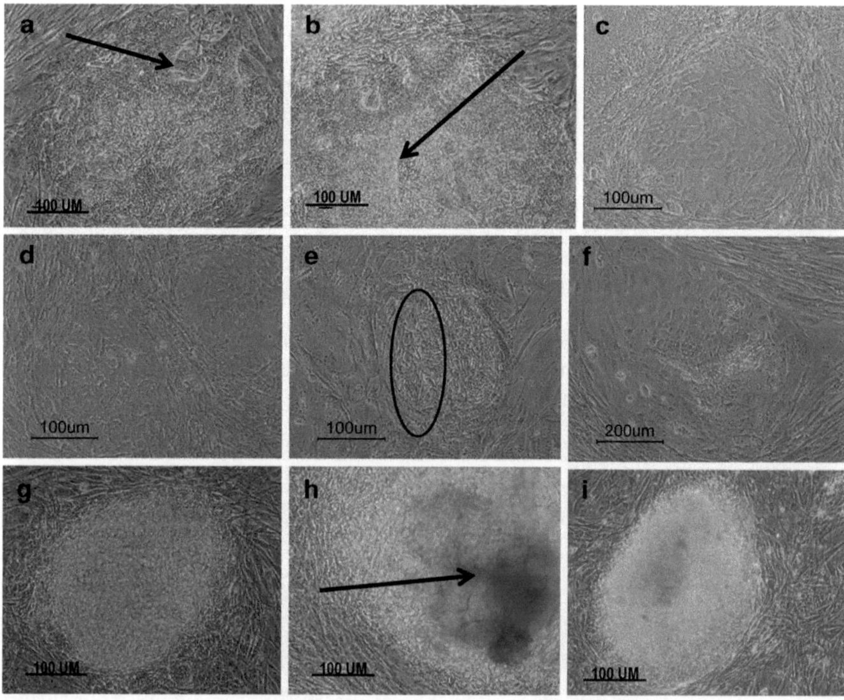

Fig. 2.23 Differentiated colonies. (**a, b, h**) Colonies in which differentiating cells formed structures (marked by *arrows*). (**c, d**) Very flat colonies with unusually large cells. Some cells may still be undifferentiated. (**e**) Partly differentiated colony; the differentiating cells are marked with a *circle*. (**f**) A colony containing large cells, with a typical nucleus–cytoplasm ratio. (**g, i**) Thick colonies with unclear cell morphology; the cells are possibly still undifferentiated. (**a–d**) hESC BG01, (**f**) hESC H9.2, and (**e, i, g, h**) hESCs I3. Bar 100 μM for all but (**b**) which has 200 μM

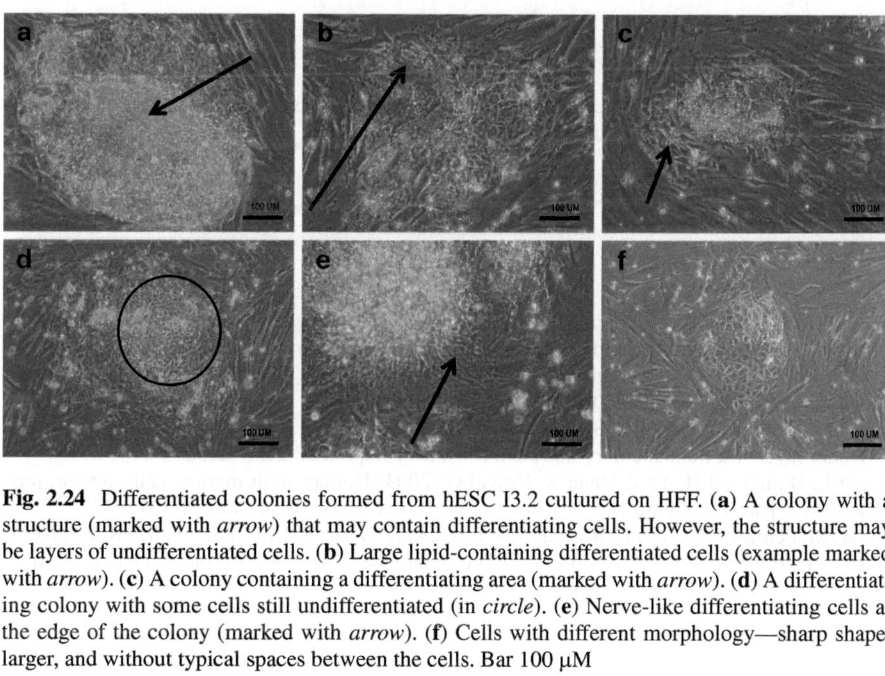

Fig. 2.24 Differentiated colonies formed from hESC I3.2 cultured on HFF. (**a**) A colony with a structure (marked with *arrow*) that may contain differentiating cells. However, the structure may be layers of undifferentiated cells. (**b**) Large lipid-containing differentiated cells (example marked with *arrow*). (**c**) A colony containing a differentiating area (marked with *arrow*). (**d**) A differentiating colony with some cells still undifferentiated (in *circle*). (**e**) Nerve-like differentiating cells at the edge of the colony (marked with *arrow*). (**f**) Cells with different morphology—sharp shape, larger, and without typical spaces between the cells. Bar 100 μM

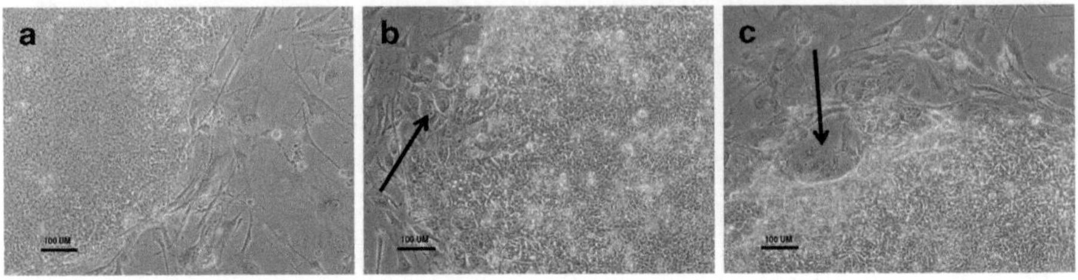

Fig. 2.25 Differentiation at the colony borders. hESC H9 colonies 7 days post-splitting (**a**). With clear border and undifferentiated morphology (**b**, **c**). With differentiating cells at the border of the colonies (marked with *arrow*). Bar 100 μM

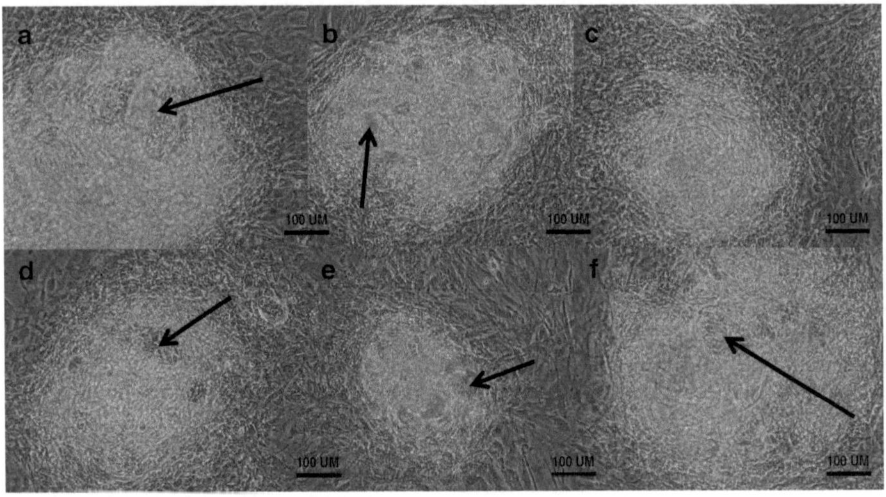

Fig. 2.26 Differentiating colonies formed from hESC I3 6 days post-splitting. All colonies formed structures within the colony (examples marked with *arrows*). Bar 100 μM

References

Aguilar-Gallardo C, Poo M, Gomez E, Galan A, Sanchez E, Marques-Mari A, Ruiz V, Medrano J, Riboldi M, Valbuena D, Simon C (2010) Derivation, characterization, differentiation, and registration of seven human embryonic stem cell lines (VAL-3, -4, -5, -6M, -7, -8, and -9) on human feeder. In Vitro Cell Dev Biol Anim 46(3–4):317–326

Amit M, Margulets V, Segev H, Shariki C, Laevsky I, Coleman R, Itskovitz-Eldor J (2003) Human feeder layers for human embryonic stem cells. Biol Reprod 68:2150–2156

Cheng L, Hammond H, Ye Z, Zhan X, Dravid G (2003) Human adult marrow cells support prolonged expansion of human embryonic stem cells in culture. Stem Cells 21:131–142

Crook JM, Peura TT, Kravets L, Bosman AG, Buzzard JJ, Horne R, Hentze H, Dunn NR, Zweigerdt R, Chua F, Upshall A, Colman A (2007) The generation of six clinical-grade human embryonic stem cell lines. Cell Stem Cell 1:490–494

Ellerström C, Strehl R, Moya K, Andersson K, Bergh C, Lundin K, Hyllner J, Semb H (2006) Derivation of a xeno-free human embryonic stem cell line. Stem Cells 24:2170–2176

Genbacev O, Krtolica A, Zdravkovic T, Brunette E, Powell S, Nath A, Caceres E, McMaster M, McDonagh S, Li Y, Mandalam R, Lebkowski J, Fisher SJ (2005) Serum-free derivation of human embryonic stem cell lines on human placental fibroblast feeders. Fertil Steril 83:1517–1529

Hovatta O, Mikkola M, Gertow K, Stromberg AM, Inzunza J, Hreinsson J, Rozell B, Blennow E, Andang M, Ahrlund-Richter L (2003) A culture system using human foreskin fibroblasts as feeder cells allows production of human embryonic stem cells. Hum Reprod 18:1404–1409

Ilic D, Giritharan G, Zdravkovic T, Caceres E, Genbacev O, Fisher SJ, Krtolica A (2009) Derivation of human embryonic stem cell lines from biopsied blastomeres on human feeders with minimal exposure to xenomaterials. Stem Cells Dev 18:1343–1350

Inzunza J, Gertow K, Strömberg MA, Matilainen E, Blennow E, Skottman H, Wolbank S, Ahrlund-Richter L, Hovatta O (2005) Derivation of human embryonic stem cell lines in serum replacement medium using postnatal human fibroblasts as feeder cells. Stem Cells 23:544–549

Richards M, Fong CY, Chan WK, Wong PC, Bongso A (2002) Human feeders support prolonged undifferentiated growth of human inner cell masses and embryonic stem cells. Nat Biotechnol 20:933–936

Simón C, Escobedo C, Valbuena D, Genbacev O, Galan A, Krtolica A, Asensi A, Sánchez E, Esplugues J, Fisher S, Pellicer A (2005) First derivation in Spain of human embryonic stem cell lines: use of long-term cryopreserved embryos and animal-free conditions. Fertil Steril 83:246–249

Ström S, Holm F, Bergström R, Strömberg AM, Hovatta O (2010) Derivation of 30 human embryonic stem cell lines. In Vitro Cell Dev Biol Anim 46(3–4):337–344

Takahashi K, Yamanaka S (2006) Induction of pluripotent stem cells from mouse embryonic and adult fibroblast cultures by defined factors. Cell 126:663–676

Takahashi K, Tanabe K, Ohnuki M, Narita M, Ichisaka T, Tomoda K, Yamanaka S (2007) Induction of pluripotent stem cells from adult human fibroblasts by defined factors. Cell 131:861–872

Tecirlioglu RT, Nguyen L, Koh K, Trounson AO, Michalska AE (2010) Derivation and maintenance of human embryonic stem cell line on human adult skin fibroblast feeder cells in serum replacement medium. In Vitro Cell Dev Biol Anim 46(3–4):231–235

Thomson JA, Itskovitz-Eldor J, Shapiro SS, Waknitz MA, Swiergiel JJ, Marshall VS, Jones JM (1998) Embryonic stem cell lines derived from human blastocysts. Science 282:1145–1147 [erratum in Science 1998;282:1827]

Valbuena D, Galán A, Sánchez E, Poo ME, Gómez E, Sánchez-Luengo S, Melguizo D, García A, Ruiz V, Moreno R, Pellicer A, Simón C (2006) Derivation and characterization of three new Spanish human embryonic stem cell lines (VAL -3 -4 -5) on human feeder and in serum-free conditions. Reprod Biomed Online 13:875–886

Ellerström C, Strehl R, Moya K, Andersson K, Bergh C, Lundin K, Hyllner J, Semb H (2006) Derivation of a xeno-free human embryonic stem cell line. Stem Cells 24:2170–2176

Genbacev O, Krtolica A, Zdravkovic T, Brunette E, Powell S, Nath A, Caceres E, McMaster M, McDonagh S, Li Y, Mandalam R, Lebkowski J, Fisher SJ (2005) Serum-free derivation of human embryonic stem cell lines on human placental fibroblast feeders. Fertil Steril 83:1517–1529

[illegible]

Chapter 3
Morphology of Human Embryonic Stem Cells and Induced Pluripotent Stem Cells Cultured in Feeder Layer-Free Conditions

Abstract While the culturing of human embryonic stem cells (hESCs) and induced pluripotent cells (iPSCs) with mouse embryonic fibroblasts (MEFs) or human foreskin fibroblasts may be adequate for research purposes, it is less suitable for clinical and industrial use due to the risk of exposure of the cells to animal pathogens and to variations between batches of MEFs and fetal bovine serum (FBS). To establish reproducible and defined cultures, devoid of animal products, feeder layers and animal products need to be replaced. This chapter describes layer-free culture systems that have been developed specifically for iPSCs.

3.1 Introduction

The traditional culture and isolation methods for human embryonic stem cells (hESCs) include inactivated mouse embryonic fibroblasts (MEFs) as feeder layers and medium supplemented with fetal bovine serum (FBS) (Thomson et al. 1998). The feeder layers play a dual role of supporting ESC proliferation and preventing ESC spontaneous differentiation by secreting factors such as leukemia inhibitory factor (LIF) (Smith et al. 1988; Williams et al. 1988). Similarly, most induced pluripotent stem cell (iPSC) line derivations include a stage in which reprogrammed cells are transferred to the MEF supporting layer (Park et al. 2008; Takahashi et al. 2007; Nakagawa et al. 2008; Yu et al. 2007). However, possible future uses of hESCs and iPSCs such as cell therapy and drug screening will require a reproducible, defined, and animal-free culture system for their derivation and routine culture. To prevent their exposure to animal photogens, human pluripotent stem cells (hPSCs) should be cultured with a medium devoid of any animal or non-defined products, and the MEFs should be replaced by human feeders or by defined matrices (synthetic if possible), using a feeder layer-free culture system for hPSC expansion.

Concerted worldwide efforts to meet these requirements have led to a number of scientific advances. Importantly, MEFs were replaced with human feeders, and

M. Amit and J. Itskovitz-Eldor, *Atlas of Human Pluripotent Stem Cells: Derivation and Culturing*, Stem Cell Biology and Regenerative Medicine,
DOI 10.1007/978-1-61779-548-0_3, © Springer Science+Business Media, LLC 2012

medium was supplemented with either human serum or serum replacement, as described in this chapter. Although animal-product free, the resultant culture conditions remain undefined, due to the use of a supporting layer. The ideal culture method for hPSCs would, therefore, combine a defined animal-free medium with a synthetic matrix.

The first feeder layer-free culture system for hESCs was reported in 2001 by Xu and colleagues (2001). The culture method, based on 100% MEF-conditioned medium supplemented with serum replacement, was combined with Matrigel, laminin, or fibronectin as the matrix (Xu et al. 2001). When cultured under these conditions, hESCs demonstrated characteristic ESC features, including stable karyotype and pluripotency, for over a year of culturing. However, this method still entailed disadvantages relating to the use of MEFs, mainly exposure to animal pathogens and possible variations between MEF batches. Several years later, the same group proposed an improvement to the culture system, whereby the MEF-conditioned medium was replaced by 40 ng/ml basic fibroblast growth factor (bFGF) and 75 ng/ml Flt-3 ligand (Xu et al. 2005a, b).

Investigators throughout the world have proposed additional improvements to the culture system. These include medium supplemented with combined transforming growth factor beta 1($TGF_{\beta 1}$) and bFGF (Amit et al. 2004), glycogen synthase kinase-3 (GSK-3)-specific inhibitor—Bio (Sato et al. 2004), Activin (Beattie et al. 2005), a high level of Noggin (suppression of BMP signaling) (Xu et al. 2005b), a high concentration of bFGF (Wang et al. 2005; Xu et al. 2005b), and a merge of five factors used in a defined culture medium (Ludwig et al. 2006). These were all found to maintain hESC features after prolonged periods of culture.

It is not yet completely known which of the sources used for reprogramming iPSC lines, or different genes and vectors involved, rely on the same signaling pathways for self-maintenance. Therefore, it is not certain that all culture systems developed for hESCs can be applied without adaptation to iPSC cultures. Preliminary results demonstrate that at least some of the culture methods developed for hESCs can also be applied to iPSCs. Vallier et al. demonstrated that similar to hESCs, human iPSCs rely on the Activin–Nodal pathway to control Nanog expression, and thereby pluripotency (Vallier et al. 2009). In that study, hiPSCs were cultured using a defined medium based on the supplement of insulin, transferring activin and FGF2, and fibronectin as the matrix. Under these conditions, the researchers were able to reprogram human foreskin fibroblast, fetal lung fibroblast, and adult dermal fibroblast, demonstrating the capability of culturing and reprogramming cells from various sources (Vallier et al. 2009).

Other studies have reported a feeder layer-free culture of iPSCs and reprogramming based on matrices, such as Matrigel, or human proteins, such as laminin, and a defined medium supplemented with growth factors, such as a high concentration of bFGF (Nagaoka et al. 2010; Rodin et al. 2010; Rodríguez-Pizà et al. 2010; Totonchi et al. 2010). Overall, it seems that similar to hESCs, a feeder layer-free culture system can be adapted for iPSC culture.

Recently, a new concept for a feeder layer-free culture of hPSCs was proposed, whereby cells are cultured in suspension either statically in petri dishes or dynamically in shaking Erlenmeyer or spinner flasks (Amit et al. 2010; Olmer et al. 2010; Singh et al. 2010; Steiner et al. 2010). The new suspension cultures are detailed in Chap. 5.

Traditionally, ESCs are isolated while using a supporting layer. Therefore, it is not surprising that the vast majority of the existing hESC lines have been isolated while using a feeder layer, mainly MEFs (Thomson et al. 1998; Reubinoff et al. 2000; Amit and Itskovitz-Eldor 2002; Cowan et al. 2004). Other types of feeder layers, such as foreskin fibroblasts, have also been used for clinical-grade hESCs (Ellerström et al. 2006; Crook et al. 2007).

Only a few reports have addressed the feeder layer-free derivation of hESCs. The first demonstrated ESC features in six new hESC lines derived from a culture system consisting of a MEF-produced matrix and a medium supplemented with a high dose of bFGF (16 ng/ml), LIF, serum replacement, and plasmanate (Klimanskaya et al. 2005). This pioneering study demonstrates the feasibility of a feeder layer-free derivation of hESCs, though the culture system included some non-defined materials.

A defined feeder-free system was reported in a study that isolated two hESC lines using a defined serum animal-free medium and a matrix containing recombinant human proteins (Ludwig et al. 2006). The lines generated under these conditions maintained hESC features after prolonged periods of culture, but displayed karyotypic abnormalities. It has yet to be determined whether the embryos were originally defective or represented exceptional events of karyotype abnormalities that occurred during the extended periods of culture, or whether this culture method does not support the derivation of hESC lines with normal karyotype.

Another study of feeder layer-free isolation of hESC lines utilized the suspension technique (Steiner et al. 2010). Isolated inner cell mass was transferred to an animal-free medium for proliferation in suspension. Three new hESC lines were derived (reported success rates were about 20%), exhibiting normal hESC features. This demonstrated that feeders or extracellular matrices are not necessarily required if suitable conditions are employed (Steiner et al. 2010).

3.2 Materials for Feeder Layer-Free Culture of hPSCs

3.2.1 Matrix Preparation

3.2.1.1 Fibronectin

50 μg/10 cm^2 fibronectin (Sigma human foreskin fibroblast cellular fibronectin F6277; Sigma human plasma fibronectin F2006; Chemicon human plasma fibronectin FC010-10) diluted with sterile water (Sigma W1503).

3.2.1.2 Matrigel

Matrigel (BD Biosciences, Bedford, MA, USA) diluted 1:40 in Dulbecco's modified Eagle's medium (DMEM) (of any type), according to the manufacturer's instructions.

3.2.2 Culture Medium

3.2.2.1 TLF Medium

84% DMEM/F12 (Biological Industries Ltd); 15% knockout (KO) serum replacement (SR, Invitrogen Corporation, C.N. 10828028); 1% nonessential amino acid (Invitrogen Corporation); 2 mM L-glutamine (Invitrogen Corporation); 0.1 mM β-mercaptoethanol (Invitrogen Corporation); 4 ng/ml human recombinant bFGF and 0.12 ng/ml human recombinant $TGF_{\beta 1}$ (both factors from R&D Biosystems); and 1,000 u/ml human recombinant LIF (PeproTec).

3.2.2.2 CM100F Medium

84% DMEM/F12 (Biological Industries Ltd); 15% KO SR (Invitrogen Corporation); 1% nonessential amino acid (Invitrogen Corporation); 2 mM L-glutamine (Invitrogen); 0.1 mM β-mercaptoethanol (Invitrogen Corporation); and 4 ng/ml human recombinant bFGF and 100 ng/ml human IL6-IL6 receptor chimera (can be replaced by 25 ng/ml IL6 and IL6 soluble receptor, both from R&D Biosystems).

3.2.2.3 Commercially Available Medium

NutriStem™ (Biological Industries Ltd)
mTeST™ (Stem Cells Technologies)

Both are ready-to-use medium, to be prepared and stored according to the manufacturer's instructions.

3.3 Methods for hPSC Feeder Layer-Free Culture

3.3.1 Preparation of Matrix-Covered Plates

3.3.1.1 Covering Plates with Fibronectin

All plates should be covered with fibronectin at least 30 min before plating of the hPSCs, as follows:

1. Cover the dish with fibronectin solution (see Sect. 3.2.1.1) to reach a concentration of 50 μg/10 cm^2, with the liquid covering the entire well surface area.
2. Leave at room temperature or in the incubator for at least 30 min. Plates covered with fibronectin can be prepared in advance (overnight).

3.3.1.2 Covering Plates with Matrigel

1. The matrix should be prepared according to the manufacturer's instructions, mainly diluted with plain DMEM under 20°C conditions, while using previously chilled pipettes and plates. It is recommended that a stock of 1:20 be prepared in advance and frozen in aliquots.
2. Cover the dish with Matrigel solution (see Sect. 3.2.1.2) to reach a dilution of 1:40, with the liquid covering the entire well surface area. We recommend 1 ml of diluted Matrigel for covering a 10 cm^2 surface area.
3. Leave at room temperature or in the incubator for at least 30 min. It is recommended that the plates not be prepared in advance.

3.3.2 Splitting, Freezing, and Thawing hPSCs

The protocols that are currently in use are identical to those described in Chap. 3 for culturing hPSCs with a supporting layer. We recommend, however, that before plating the cells with matrix, residue of liquids be removed from the culture dish, after either splitting or thawing. In addition, since cultured cells are more sensitive, the use of Rho kinase (ROCK) inhibitor Y-27632 while freezing the cells (see for protocol Watanabe et al. 2007) is advised. Colonies propagated without feeder layers are flatter than those grown on feeder layers. Therefore, breaking them for the purpose of splitting or freezing may result in smaller clumps (Fig. 3.1). Another difference is that when the colonies are broken, either for splitting or freezing, the clumps may be smaller (Fig. 3.1), since the colonies tend to be flatter than when using feeders.

3.3.3 Adaptation of PSCs to Feeder-Free Culture

Usually, cells can be transferred directly from a feeder-containing culture to a feeder-free one. However, they will require 3–6 passages to adapt to the feeder layer-free culture system. During adaptation, some residues of the feeder may still exist in the culture, and the colony morphology may be disorganized (Fig. 3.2). If the cell line does not adapt well to the new culture conditions, it may help to adapt the cells to the new culture medium while still culturing with feeders, using gradual transfer: one passage with 75% old medium and 25% new medium, followed by a passage with 50%:50% medium ratio, and a passage with 75%:25%. In subsequent passages, 100% of the new medium can be used. After 2 passages with the new medium, the cells can be transferred to feeder layer-free culture conditions more easily. Similarly, transfer between medium types during feeder-free culture can be performed gradually, as described.

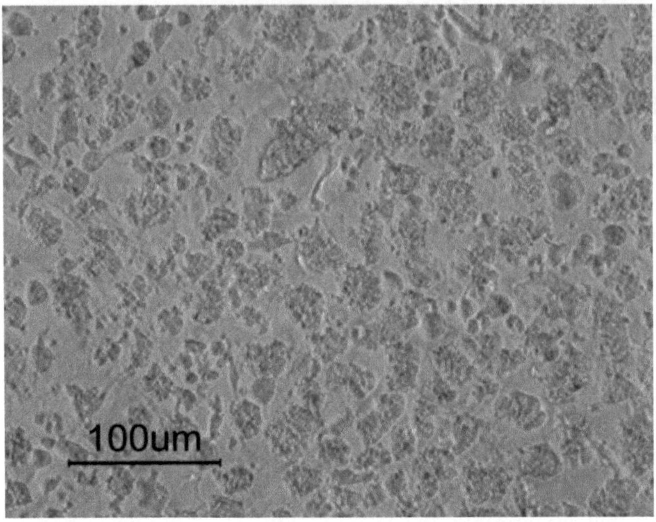

Fig. 3.1 Size of clumps post-splitting. BG01 cell clumps following collagenase splitting. They are smaller than would be expected from using feeder containing culture. Bar 100 μM

3.3.4 Routine Culture of hPSCs

The culture medium (see Sect. 3.2.2) should be changed daily. If this is unfeasible, add twice the amount of medium. hPSCs should be passaged every 4–6 days directly on fresh fibronectin or Matrigel-covered plates (see Sects. 3.3.1.1 or 3.3.1.2). Differentiating colonies should be scraped after every 5–7 passages. The undifferentiated colony morphology resembles those of colonies cultured with feeders, consisting of small round cells with large nuclei and notable nucleoli, and typical spaces between the cells. The major differences are that the colonies are sometimes less rounded and with sharp edges, and are flatter, comprising fewer layers. Examples of undifferentiated morphology are illustrated in Figs. 3.3–3.6 and 3.8a for TLF, CM100F, NutriStem™, mTeSR™, and TeSR2™ media, respectively. In some cases, due to the flat morphology at the end of the passages, the undifferentiated cells may form a monolayer (Fig. 3.7). As with feeder containing culture systems, background differentiation is retained in the PSC culture. One type of background differentiation is the formation of auto-feeders from PSCs (described by Xu et al. 2001). These auto-feeders remain during collagenase splitting and re-form at every passage. A culture containing auto-feeders may become clean during prolonged culture, yet the appearance of auto-feeders is hard to prevent. High bFGF concentrations of 40 ng/ml or more have been recommended

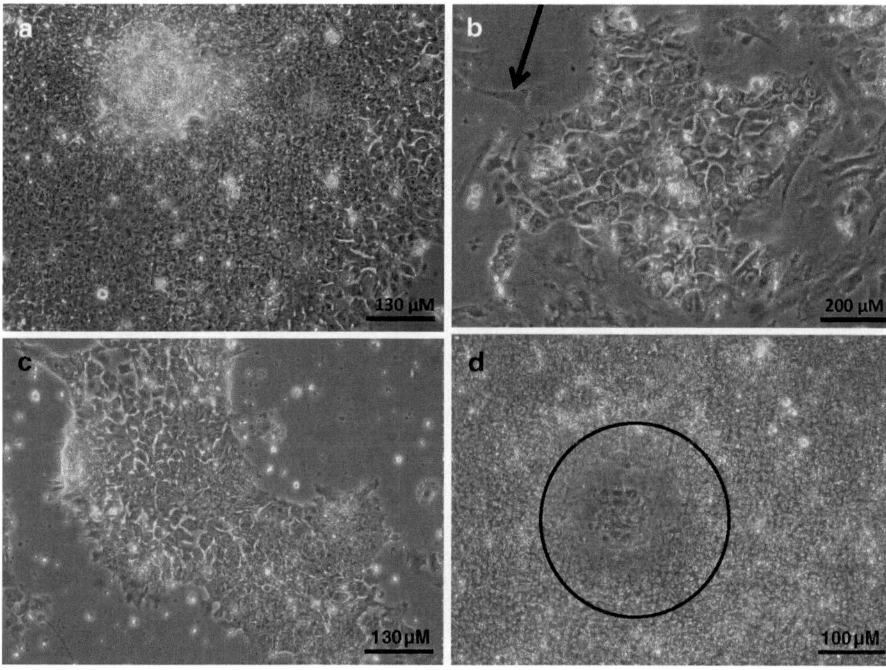

Fig. 3.2 Colonies adapting to feeder layer-free culture, cultivated on Matrigel™ matrix, demonstrating disorganized morphology. (**a**) A colony of hESCs CL1 cultured with mTeSR™ containing undifferentiated cells. (**b**) A colony of hESCs CL1 cultured with mTeSR™ containing undifferentiated-like cells with spaces between them. The cell size is larger than usual, and the colony is not round. It seems that there are still some feeder cells at the culture (an example is marked with an *arrow*). (**c**) A colony of hESCs I3.2 cultured with NutriStem™ with elongated shape. (**d**) A large colony of hESCs CL1 mainly containing undifferentiated cells. At the colony center, there is an area (marked with a *circle*) where the cell morphology is unclear. (**a**, **c**) Bar 130 μM, (**b**) bar 200 μM, and (**d**) bar 100 μM

as a means of preventing auto-feeder formation (Ludwig et al. 2006). Figure 3.6 shows undifferentiated cells forming auto-feeders. Background differentiation is also characteristic of the feeder-free culture, with similar appearance as that seen in feeder-based cultures (Figs. 3.8 and 3.9).

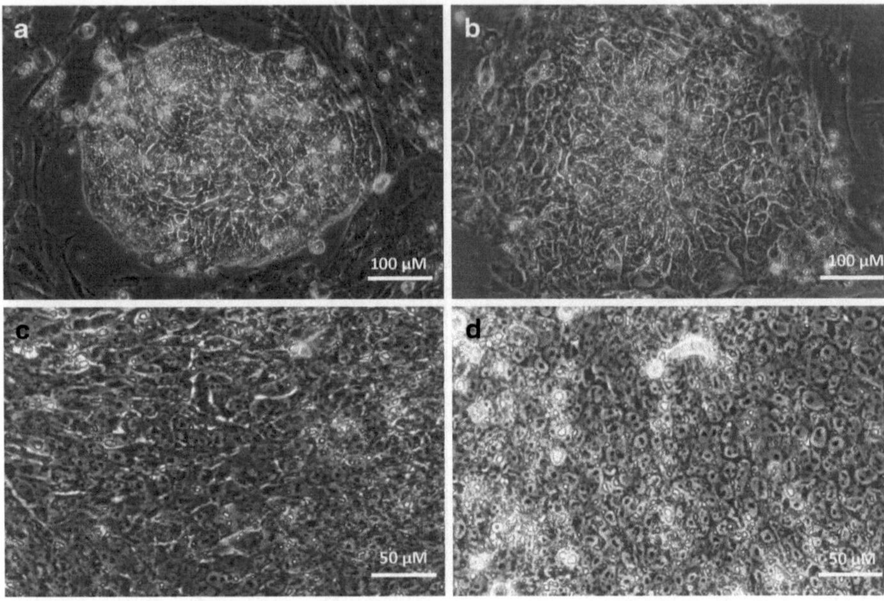

Fig. 3.3 Undifferentiated hESC I3 cultured with TLF medium and fibronectin. (**a**, **b**) Examples of undifferentiated cell colonies. Note that some auto-feeders surround the colonies. (**c**, **d**) The cell morphology of colonies is identical to the one expected in cells cultured on feeder cells. (**a**, **b**) Bar 100 μM and (**c**, **d**) bar 50 μM

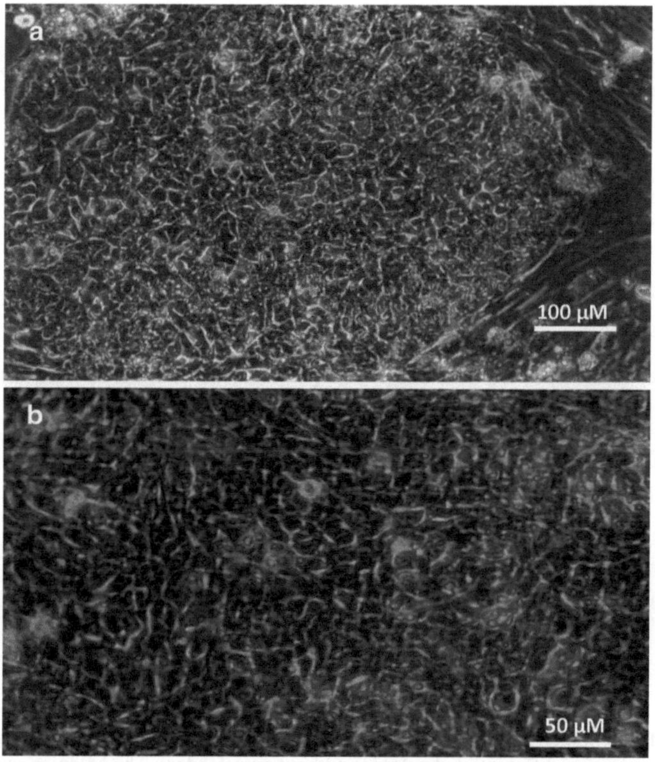

Fig. 3.4 Undifferentiated hESC I3 cultured with CM100F medium and fibronectin. (**a**) An example of an undifferentiated cell colony. Note the auto-feeders surrounding it. (**b**) A typical cell morphology of a feeder layer-free culture colony. (**a**) Bar 100 µM and (**b**) bar 50 µM

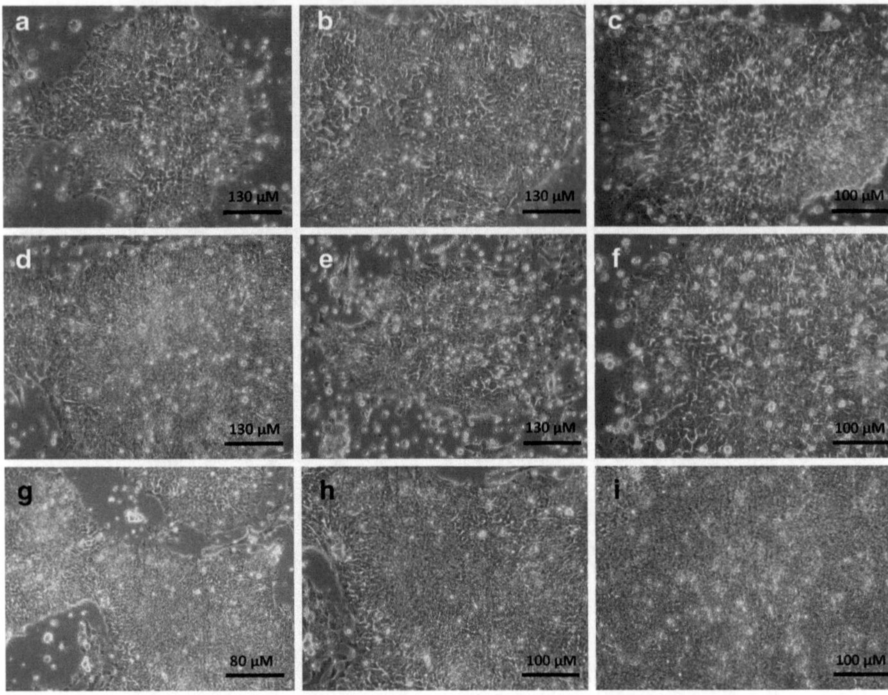

Fig. 3.5 Undifferentiated hESC I3.2 cultured with NutriStem™ medium and Matrigel™ matrix. (**a–h**) An example of undifferentiated colonies without auto-feeders. The colonies are not round as observed in feeder cell-based cultures, and have sharp edges. The culture is accompanied by floating dead cells. (**i**) Morphology of undifferentiated CL1 hESC. The colony is larger than expected for cells cultured with feeder cells, almost creating a monolayer. No signs of cell differentiation. (**a**, **b**, **d**, **e**) Bar 130 μM, (**c**, **f**, **h**, **i**) bar 100 μM, and (**g**) bar 80 μM

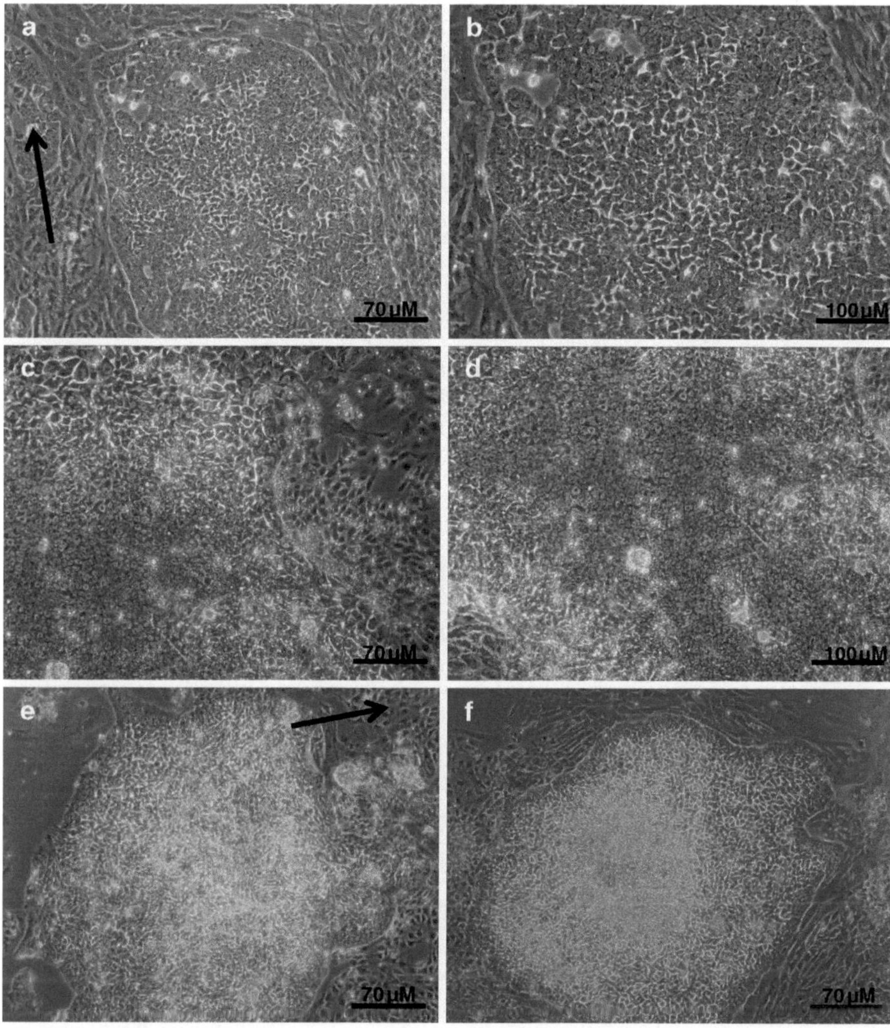

Fig. 3.6 Undifferentiated hESC CL1 cultured with Matrigel™ matrix, accompanied with auto-feeders. (**a**) An undifferentiated colony cultured with NutriStem™, surrounded by auto-feeder cells and a small colony (marked with *arrow*), in which (**b**) no clear differentiation is apparent within the colony. (**c**) An undifferentiated colony cultured with mTeSR™, surrounded with auto-feeders, in which (**d**) no clear differentiation is apparent within the colony. (**e**) An undifferentiated colony cultured with mTeSR™, in which auto-feeders are formed only on one side of the colony (*arrow*), whereas the colony in figure (**f**) is completely surrounded by auto-feeders. Both colonies designate clear borders that separate them from the auto-feeders. (**a**, **c**, **e**, **f**) Bar 70 µM and (**b**, **d**) bar 100 µM

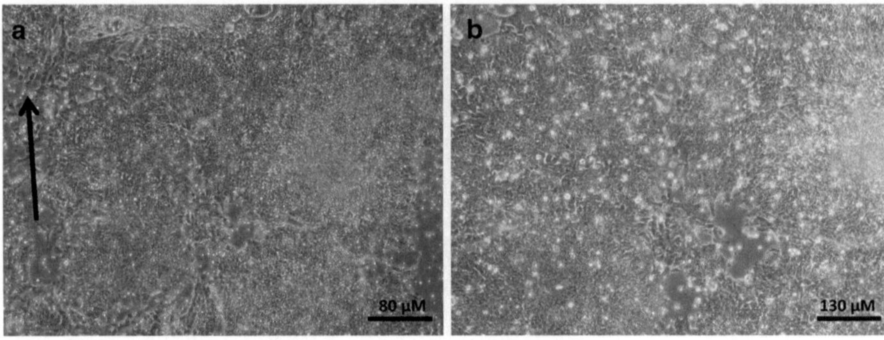

Fig. 3.7 Undifferentiated hESC I3.2 cultured with Matrigel™ matrix and NutriStem™, creating a monolayer. In (**a**), an area with some auto-feeders is apparent (marked with *arrow*), while (**b**) consists of undifferentiated cells only. (**a**) Bar 80 μM and (**b**) bar 130 μM

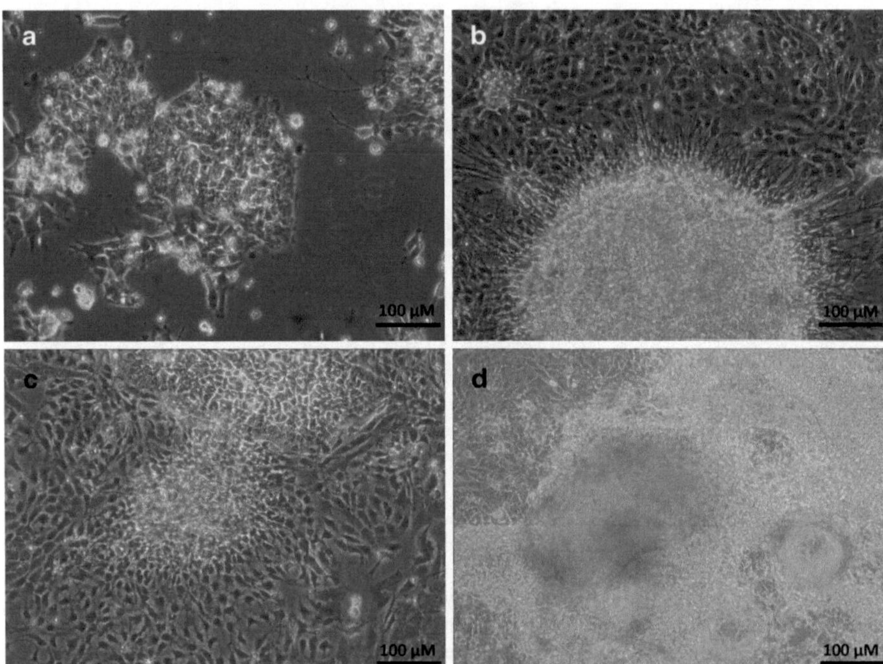

Fig. 3.8 Differentiated hESC I3.2 cultured with Matrigel™ matrix and TeSR2™. (**a**) Small undifferentiated colonies. (**b**) A differentiated colony containing cells with neuron-like morphology at the edges. (**c**) Flat differentiated colonies and (**d**) a thick colony containing structured differentiated cells. Bar 100 μM

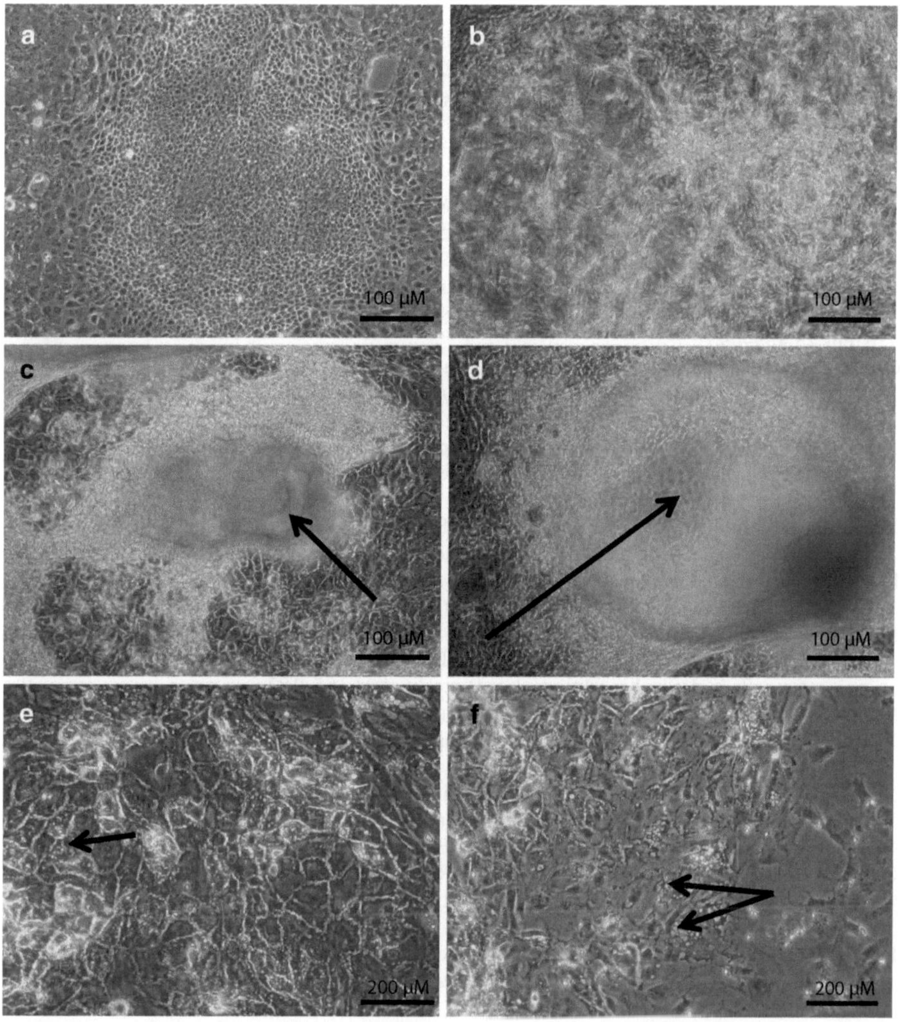

Fig. 3.9 Differentiated hESC CL1 cultured with Matrigel™ matrix. (**a**) A flat colony cultured with NutriStem™, containing a dark cell with no notable nucleus. (**b**) Differentiated colony cultured with NutriStem™, containing tube-like structures. (**c**) A large colony cultured with mTeSR™, containing epithelium-like structure (marked with *arrow*). (**d**) A large colony cultured with mTeSR™, with a cyst formation (marked with *arrow*). (**e–f**) Colonies cultured with mTeSR™, containing large cells with lipid droplet (marked with *arrow*). (**a–d**) Bar 100 μM and (**e**, **f**) bar 200 μM

References

Amit M, Itskovitz-Eldor J (2002) Derivation and spontaneous differentiation of human embryonic stem cells. J Anat 200:225–232

Amit M, Shariki K, Margulets V, Itskovitz-Eldor J (2004) Feeder and serum-free culture system for human embryonic stem cells. Biol Reprod 70:837–845

Amit M, Chebath J, Marguletz V, Laevsky I, Miropolsky Y, Shariki K, Peri M, Revel M, Itskovitz-Eldor J (2010) Suspension culture of undifferentiated human embryonic and induced pluripotent stem cells. Stem Cells Rev Rep 6:248–259

Beattie GM, Lopez AD, Bucay N, Hinton A, Firpo MT, King CC, Hayek A (2005) Activin A maintains pluripotency of human embryonic stem cells in the absence of feeder layers. Stem Cells 23:489–495

Cowan CA, Klimanskaya I, McMahon J, Atienza J, Witmyer J, Zucker JP, Wang S, Morton CC, McMahon AP, Powers D, Melton DA (2004) Derivation of embryonic stem-cell lines from human blastocysts. N Engl J Med 350:1353–1356

Crook JM, Peura TT, Kravets L, Bosman AG, Buzzard JJ, Horne R, Hentze H, Dunn NR, Zweigerdt R, Chua F, Upshall A, Colman A (2007) The generation of six clinical-grade human embryonic stem cell lines. Cell Stem Cell 1:490–494

Ellerström C, Strehl R, Moya K, Andersson K, Bergh C, Lundin K, Hyllner J, Semb H (2006) Derivation of a xeno-free human embryonic stem cell line. Stem Cells 24:2170–2176

Klimanskaya I, Chung Y, Meisner L, Johnson J, West MD, Lanza R (2005) Human embryonic stem cells derived without feeder cells. Lancet 365:1636–1641

Ludwig TE, Levenstein ME, Jones JM, Berggren WT, Mitchen ER, Frane JL, Crandall LJ, Daigh CA, Conard KR, Piekarczyk MS, Llanas RA, Thomson JA (2006) Derivation of human embryonic stem cells in defined conditions. Nat Biotechnol 24:185–187

Nagaoka M, Si-Tayeb K, Akaike T, Duncan SA (2010) Culture of human pluripotent stem cells using completely defined conditions on a recombinant E-cadherin substratum. BMC Dev Biol 2:10–60

Nakagawa M, Koyanagi M, Tanabe K, Takahashi K, Ichisaka T, Aoi T, Okita K, Mochiduki Y, Takizawa N (2008) Yamanaka S (2008) Generation of induced pluripotent stem cells without Myc from mouse and human fibroblasts. Nat Biotechnol 26:101–106

Olmer R, Haase A, Merkert S, Cui W, Palecek J, Ran C, Kirschning A, Scheper T, Glage S, Miller K, Curnow EC, Hayes ES, Martin U (2010) Long term expansion of undifferentiated human iPS and ES cells in suspension culture using a defined medium. Stem Cell Res 5:51–64

Park IH, Zhao R, West JA, Yabuuchi A, Huo H, Ince TA, Lerou PH, Lensch MW, Daley GQ (2008) Reprogramming of human somatic cells to pluripotency with defined factors. Nature 451:141–146

Reubinoff BE, Pera MF, Fong C, Trounson A, Bongso A (2000) Embryonic stem cell lines from human blastocysts: somatic differentiation *in vitro*. Nat Biotechnol 18:399–404

Rodin S, Domogatskaya A, Ström S, Hansson EM, Chien KR, Inzunza J, Hovatta O, Tryggvason K (2010) Long-term self-renewal of human pluripotent stem cells on human recombinant laminin-511. Nat Biotechnol 28:611–615

Rodríguez-Pizà I, Richaud-Patin Y, Vassena R, González F, Barrero MJ, Veiga A, Raya A, Belmonte JC (2010) Reprogramming of human fibroblasts to induced pluripotent stem cells under xeno-free conditions. Stem Cells 28:36–44

Sato N, Meijer L, Skaltsounis L, Greengard P, Brivanlou AH (2004) Maintenance of pluripotency in human and mouse embryonic stem cells through activation of Wnt signaling by a pharmacological GSK-3-specific inhibitor. Nat Med 10:55–63

Singh H, Mok P, Balakrishnan T, Rahmat SN, Zweigerdt R (2010) Up-scaling single cell-inoculated suspension culture of human embryonic stem cells. Stem Cell Res 4:165–179

Smith AG, Heath JK, Donaldson DD, Wong GG, Moreau J, Stahl M, Rogers D (1988) Inhibition of pluripotential embryonic stem cell differentiation by purified polypeptides. Nature 336: 688–690

Steiner D, Khaner H, Cohen M, Even-Ram S, Gil Y, Itsykson P, Turetsky T, Idelson M, Aizenman E, Ram R, Berman-Zaken Y, Reubinoff B (2010) Derivation, propagation and controlled differentiation of human embryonic stem cells in suspension. Nat Biotechnol 28:361–364

Takahashi K, Tanabe K, Ohnuki M, Narita M, Ichisaka T, Tomoda K, Yamanaka S (2007) Induction of pluripotent stem cells from adult human fibroblasts by defined factors. Cell 131(5): 861–872

Thomson JA, Itskovitz-Eldor J, Shapiro SS, Waknitz MA, Swiergiel JJ, Marshall VS, Jones JM (1998) Embryonic stem cell lines derived from human blastocysts. Science 282:1145–1147 [erratum in Science (1998) 282, 1827]

Totonchi M, Taei A, Seifinejad A, Tabebordbar M, Rassouli H, Farrokhi A, Gourabi H, Aghdami N, Hosseini-Salekdeh G, Baharvand H (2010) Feeder- and serum-free establishment and expansion of human induced pluripotent stem cells. Int J Dev Biol 54:877–886

Vallier L, Touboul T, Brown S, Cho C, Bilican B, Alexander M, Cedervall J, Chandran S, Ahrlund-Richter L, Weber A, Pedersen RA (2009) Signaling pathways controlling pluripotency and early cell fate decisions of human induced pluripotent stem cells. Stem Cells 27:2655–2666

Wang L, Li L, Menendez P, Cerdan C, Bhatia M (2005) Human embryonic stem cells maintained in the absence of mouse embryonic fibroblasts or conditioned media are capable of hematopoietic development. Blood 105:4598–4603

Watanabe K, Ueno M, Kamiya D, Nishiyama A, Matsumura M, Wataya T, Takahashi JB, Nishikawa S, Nishikawa S, Muguruma K, Sasai Y (2007) ROCK inhibitor permits survival of dissociated human embryonic stem cells. Nat Biotechnol 25:681–686

Williams R, Hilton D, Pease S, Wilson T, Stewart C, Gearing D, Wagner E, Metcalf D, Nicola N, Gough N (1988) Myeloid leukemia inhibitory factor maintains the developmental potential of embryonic stem cells. Nature 336:684–687

Xu C, Inokuma MS, Denham J, Golds K, Kundu P, Gold JD, Carpenter MK (2001) Feeder-free growth of undifferentiated human embryonic stem cells. Nat Biotechnol 19:971–974

Xu C, Rosler E, Jiang J, Lebkowski JS, Gold JD, O'Sullivan C, Delavan-Boorsma K, Mok M, Bronstein A, Carpenter MK (2005a) Basic fibroblast growth factor supports undifferentiated human embryonic stem cell growth without conditioned medium. Stem Cells 23:315–323

Xu RH, Peck RM, Li DS, Feng X, Ludwig T, Thomson JA (2005b) Basic FGF and suppression of BMP signaling sustain undifferentiated proliferation of human ES cells. Nat Methods 2: 185–190

Yu J, et al. (2007) Induced pluripotent stem cell lines derived from human somatic cells. Science 318:1917–1920

Chapter 4
Morphology of Undifferentiated Human Embryonic and Induced Stem Cells Grown in Suspension and in Dynamic Cultures

Abstract The realization of the potential in research and therapy of human pluripotent stem cells (hPSCs), both embryonic and induced, will require mass production of undifferentiated and differentiated cells. Clearly, the conventional two-dimensional techniques for hPSCs are inadequate for such. This chapter focuses on suspension culturing of undifferentiated hPSCs, a method suitable for the robust production of these cells in controlled bioreactors.

4.1 Introduction

Human pluripotent stem cells (hPSCs), both embryonic and induced, are usually derived and cultured as adherent cells, using feeder layers such as mouse embryonic fibroblasts (MEFs) or foreskin fibroblasts (HFFs), with medium supplemented by serum replacement and basic fibroblast growth factor (bFGF) (Amit et al. 2000, 2003). Improvements in the basic adherent culture systems are discussed in Chaps. 2 and 3.

Possible future uses of hPSCs for cell-based therapy and industrial purposes, such as drug screening, will require a reproducible, defined, and animal-free culture system for the routine culture of undifferentiated cells and for new cell line derivation. Controllable large-scale bioprocesses are necessary for the mass production of cells. Such processes generally rely on suspension culture in stirred systems, in contrast to the traditional culture of hPSCs, which is based on two-dimensional culture in culture dishes. When transferred into suspension culture in vitro, under standard growth conditions, hPSCs spontaneously form cell aggregates known as embryoid bodies (EBs), in which cells differentiate into embryonically distinct cell types (Doetschman et al. 1985; Itskovitz-Eldor et al. 2000; Takahashi et al. 2007; Yu et al. 2007).

A recently developed method for suspension culture enables the culturing of hPSCs as undifferentiated cells on coated beads, mimicking the adherent culture

M. Amit and J. Itskovitz-Eldor, *Atlas of Human Pluripotent Stem Cells: Derivation and Culturing*, Stem Cell Biology and Regenerative Medicine,
DOI 10.1007/978-1-61779-548-0_4, © Springer Science+Business Media, LLC 2012

environment (Phillips et al. 2008; Oh et al. 2009). However, uploading and removing cells from beads at each passage are challenging, and may result in cell loss (Phillips et al. 2008; Oh et al. 2009). A novel technique of suspension culturing avoids the use of microbeads; cells are propagated as small spheres in static cultures in petri dishes or in dynamic systems with shaking Erlenmeyer or spinner flasks (Amit et al. 2010; Olmer et al. 2010; Singh et al. 2010; Steiner et al. 2010).

The system described in this chapter is based on a serum-free medium, enabling a considerable increase in viable cell numbers, of 26-fold in 10 days (Amit et al. 2010, 2011). Seven hESC lines and six iPSC lines from different sources (foreskin fibroblasts, adult skin fibroblasts, and keratinocytes) were adapted to this system, using serum-free medium supplemented with full-length IL6RIL6 chimera and low concentrations of bFGF. The system demonstrated expansion for over 40 passages while maintaining pluripotency, as evidenced by teratoma and EB formation, as well as by stable karyotypes. This is the first description of a highly efficient method for the continuous long-term expansion of undifferentiated hESCs in 3D and dynamic cultures that is applicable to the large-scale cell production of hPSCs.

4.2 Materials for Suspension Culture of hPSCs

4.2.1 Culture Medium

4.2.1.1 Supplemented with IL6-IL6 Receptor Chimera

84% Dulbecco's modified Eagle's medium (DMEM)/F12 (Biological Industries Ltd); 15% knockout (KO) serum replacement (SR, Invitrogen Corporation, C.N. 10828028); 1% nonessential amino acid (Invitrogen Corporation); 2 mM L-glutamine (Invitrogen Corporation); 0.1 mM β-mercaptoethanol (Invitrogen Corporation); 4 ng/ml human recombinant bFGF; and 100 pg/ml human recombinant full-length IL6RIL6 chimera (Merck-Serono).

4.2.1.2 Supplemented with IL6 and IL6 Soluble Receptor

84% DMEM/F12 (Biological Industries Ltd); 15% KO serum replacement (SR, Invitrogen Corporation, C.N. 10828028); 1% nonessential amino acid (Invitrogen Corporation); 2 mM L-glutamine (Invitrogen Corporation); 0.1 mM β-mercaptoethanol (Invitrogen Corporation); 4 ng/ml human recombinant bFGF; and 25 ng/ml human recombinant IL6 and 25 ng/ml human recombinant IL6 soluble receptor (both from R&D BioSystems).

4.2.1.3 Supplemented with Leukemia Inhibitory Factor

85% DMEM/F12 (Biological Industries Ltd); 15% KO serum replacement (SR, Invitrogen Corporation, C.N. 10828028); 1% nonessential amino acid (Invitrogen Corporation); 2 mM L-glutamine (Invitrogen Corporation); 0.1 mM β-mercaptoethanol (Invitrogen Corporation); 4 ng/ml human recombinant bFGF; and 3,000 u/ml human recombinant leukemia inhibitory factor (LIF) (PeproTec).

4.2.2 Splitting Medium

4.2.2.1 Collagenase-Based Splitting Medium

The splitting medium consists of 1 mg/ml collagenase type IV (Wordington, type IV C.N 4189, activity of 220–320 unite/mg) in DMEM (of any type).

4.2.2.2 Trypsin-Based Splitting Medium

The splitting medium consists of trypsin–EDTA (0.25%, Biological Industries Ltd). When trypsin is used for splitting, 10 μM of ROCK inhibitor (Sigma) should be added to the cell culture medium (see Sect. 4.2.1.1 or 4.2.1.2 or 4.2.1.3) 1 h before splitting, and to the trypsin while incubating, and be left until the following morning (i.e., overnight).

4.2.3 Freezing Medium

60% DMEM (of any type), 20% dimethyl sulfoxide (DMSO) (Sigma), and 20% defined fetal bovine serum (FBS) (HyClone) or 30% knockout serum replacement (Invitrogen Corporation).

4.3 Methods for Suspension Culture of hPSCs

4.3.1 Creating a hPSC Suspension Culture

For initiating a suspension culture, it is recommended to use 5×10^6 undifferentiated hPSCs grown in adherent cultures. The cells should be split as usual (see Chap. 3). In case collagenase is used, the following protocol is recommended:

1. Aspirate the medium that is to be transferred from the wells to the suspension culture.
2. Add the splitting medium (see Sect. 4.2.2.1) to cover the well (0.5 ml for 10 cm^2) and incubate for 20–40 min. Most colonies will float.
3. Add 1 ml of culture medium (see Sect. 4.2.1.1, 4.2.1.2 or 4.2.1.3) and carefully collect the cells.
4. Collect the cell suspension and place into a conical tube.
5. Centrifuge for 3 min at $80 \times g$ at a recommended temperature of 4°C.
6. Aspirate the medium from the tube without harming the cell pellet.
7. Re-suspend the cells in 700 μl of fresh medium (see Sect. 4.2.1.1 or 4.2.1.2 or 4.2.1.3) and dissociate the cells by moving a 200-μl tip up and down 5–10 times.
8. Transfer the cells to a 58-mm dish (use a petri dish or any other non-culture-treated dish).
9. Add 5 ml of fresh medium (see Sect. 4.2.1.1, 4.2.1.2 or 4.2.1.3) and incubate at 37°C.

The morphology of undifferentiated cells cultured in suspension is shown in Fig. 4.1 for human PSCs, and in Fig. 4.2 for primate and mouse PSCs.

4.3.2 Splitting hPSCs in Suspension

4.3.2.1 Mechanical Splitting

1. Collect the cells cultured in suspension and place in a conical tube.
2. Centrifuge for 3 min at $80 \times g$ at a recommended temperature of 4°C.
3. Aspirate the medium from the tube without harming the cell pellet.
4. Re-suspend the cells in 700 μl of fresh medium (see Sect. 4.2.1.1, 4.2.1.2 or 4.2.1.3) and dissociate the cells by moving a 200-μl tip up and down 5–10 times.
5. Transfer the cells to a new 58-mm dish (use a petri dish or any other non-culture-treated dish).
6. Add 5 ml of fresh medium (see Sect. 4.2.1.1. 4.2.1.2 or 4.2.1.3) and incubate at 37°C. A concentration of 1×10^6 cells/ml is recommended.

The morphology of spheres after mechanical splitting is demonstrated in Fig. 4.3.

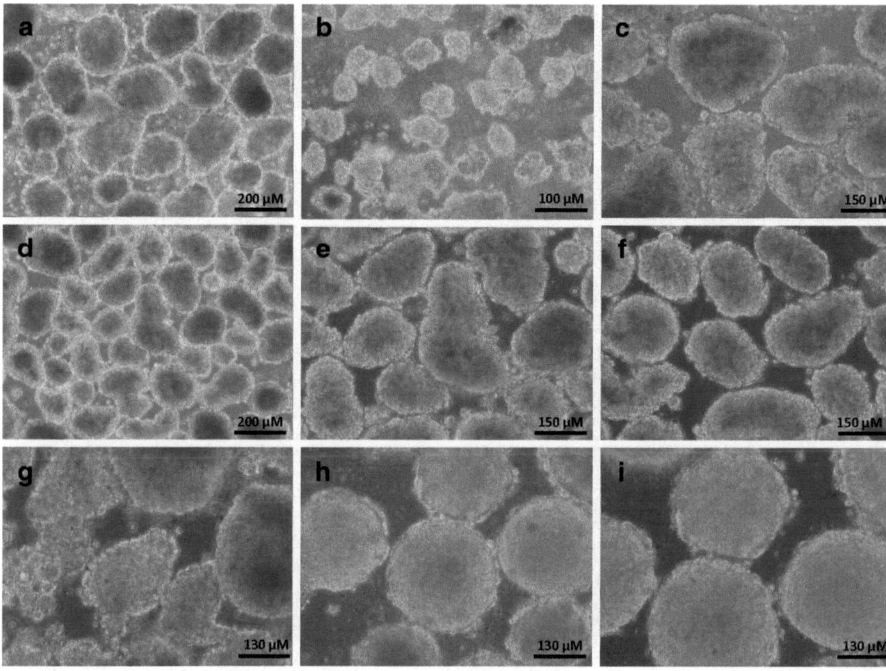

Fig. 4.1 Undifferentiated hPSCs cultured in suspension. The cells spontaneously form spheres with a diameter of 100–500 µM. (**a**) hESC line I6 cultured in suspension for 29 passages. (**b**) hESC line CL1 cultured in suspension for 6 passages. (**c**) hESC line I4 cultured in suspension for 37 passages. (**d–f**) hESC line I3 cultured in suspension for 43 passages. (**g**) Human induced pluripotent stem cell (hiPSC) line mKTR13.4 derived from plucked human hair follicle keratinocytes cultured in suspension for 5 passages. (**h, i**) hESC line H9.2 cultured in suspension for 34 passages. The formed spheres have a disc-like rather than a ball-like structure as formed by EBs, and have sharp edges. (**a, d**) Bar 200 µM, (**b**) bar 100 µM, (**c, e, f**) bar 150 µM, and (**g–i**) bar 130 µM

4.3.2.2 Splitting Cells with Trypsin

Splitting cells with trypsin is optional and will result in a more homogenous sphere size. However, the attachment after 48 h of some of the formed clumps should be considered. Consequently, the resulting spheres will not initiate from single cells.

1. Collect the cells cultured in suspension and place in a conical tube.
2. Centrifuge for 3 min at $80 \times g$ at a recommended temperature of 4°C.
3. Aspirate the medium from the tube without harming the cell pellet.
4. Add 1 ml of trypsin (see Sect. 4.2.2.2) and incubate at 37°C for 10 min.
5. Centrifuge for 3 min at $80 \times g$ at a recommended temperature of 4°C.
6. Re-suspend the cells in 700 µl of fresh medium (see Sect. 4.2.1.1, 4.2.1.2 or 4.2.1.3) supplemented with ROCK inhibitor (see Sect. 4.2.2.2), and break the cells further by moving a 200-µl tip up and down 3–5 times.

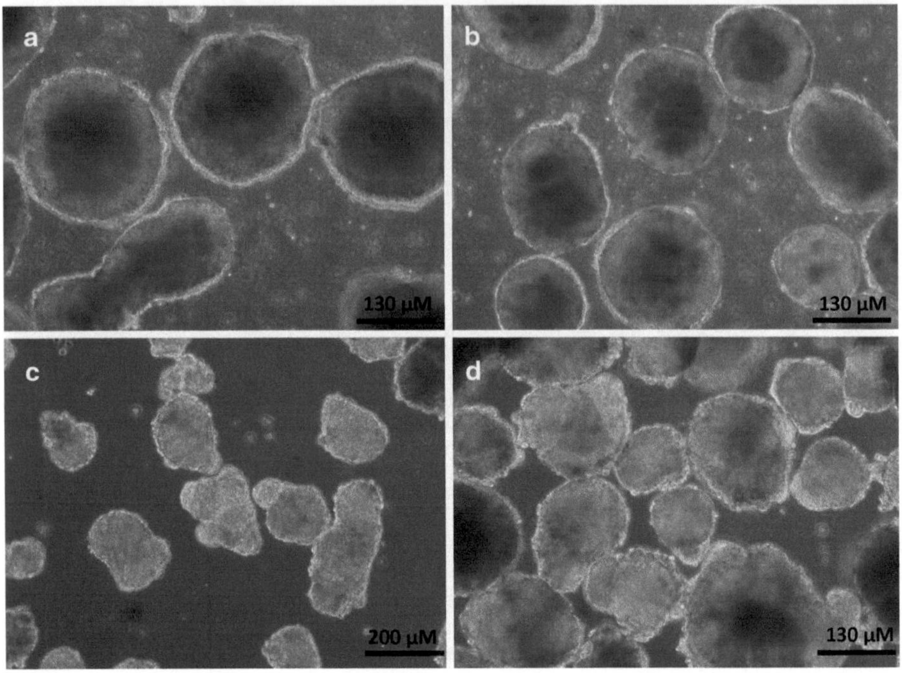

Fig. 4.2 Undifferentiated PSCs of primate and mouse origin, cultured in suspension. The sphere morphology resembles that of human spheres. (**a, b**) Rhesus monkey ESC line R366.4 cultured in suspension for 2 passages. (**c, d**) Mouse iPSC line M16 cultured in suspension for 13 passages. (**a, b, d**) Bar 130 μM and (**c**) bar 200 μM

7. Transfer the cells to a new 58-mm dish (use a petri dish or any other non-culture-treated dish). A concentration of 1×10^6 cells/ml is recommended.
8. Add 5 ml of fresh medium (see Sect. 4.2.1.1, 4.2.1.2 or 4.2.1.3) supplemented with ROCK (see Sect. 4.2.2.2) and incubate at 37°C.

   Sphere morphology after passaging with trypsin is illustrated in Fig. 4.4.

4.3.3 Freezing hPSCs in Suspension

Use the same method as that used for splitting (see Sect. 4.3.2.1 or 4.3.2.2). Freeze the cells 1 day post-splitting.

1. Collect the cells cultured in suspension and place in a conical tube.
2. Centrifuge for 3 min at $80 \times g$ at a recommended temperature of 4°C.
3. Aspirate the medium from the tube without harming the cell pellet.
4. Re-suspend the cells in a culture medium (see Sect. 4.2.1.1, 4.2.1.2 or 4.2.1.3).

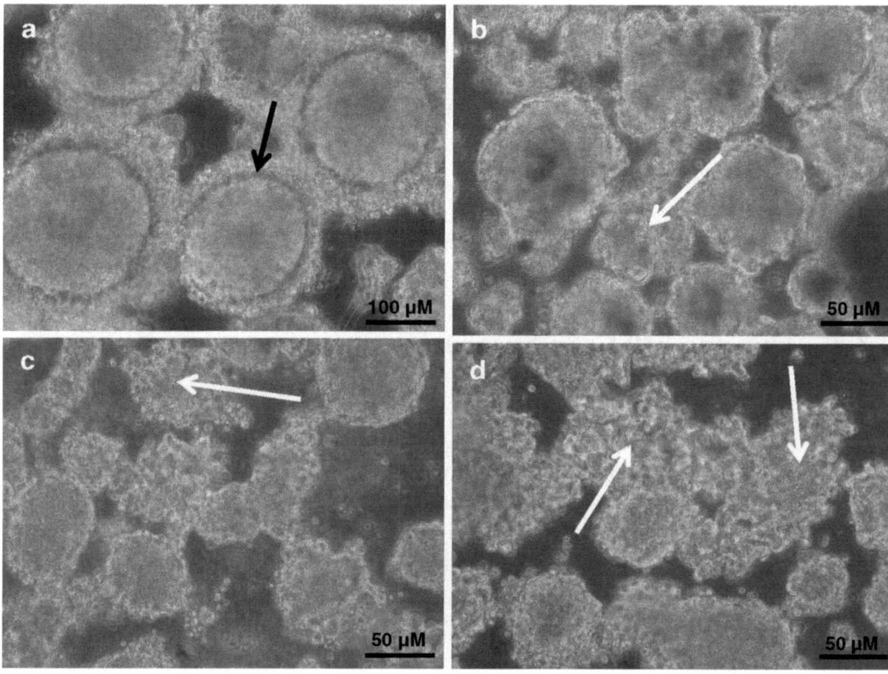

Fig. 4.3 Undifferentiated hPSCs cultured in suspension post-mechanical splitting. (**a**) hESC line I3 cultured in suspension for 2 passages. Note that the spheres already have a clear sharp border (marked with *black arrow*) surrounded by loosely attached cells. (**b**) hiPSC line mKTR13.4 cultured in suspension for 5 days. Some of the clumps do not have an organized sphere morphology (marked with *white arrow*). (**c, d**) hESC line CL1 cultured in suspension for 6 passages. Some of the clumps do not have an organized sphere morphology (marked with *white arrow*). (**a**) Bar 100 μM and (**b–d**) bar 50 μM

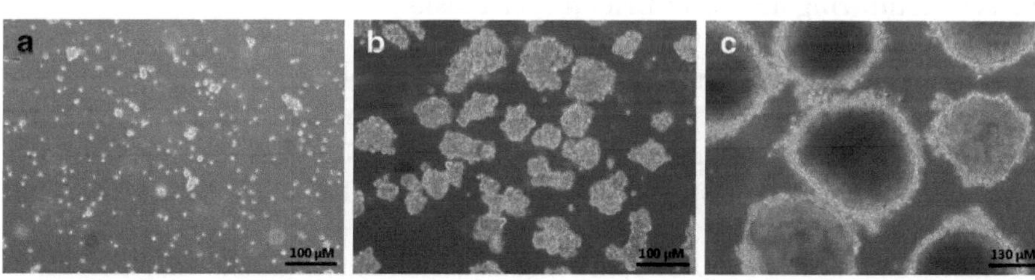

Fig. 4.4 Enzymatic splitting of the spheres. Cells of hESC line I4 that were cultured in suspension for 50 passages were dissociated with trypsin–EDTA and treated with ROCK inhibitor. One (**a**), three (**b**), and five (**c**) days after dissociation. Five days post-splitting, spheres emerged with typical undifferentiated morphology, similar to the morphology created after mechanical dissociation. (**a, b**) Bar 100 μM and (**c**) bar 130 μM

5. Add, drop by drop, an equivalent volume of freezing medium (see Sect. 5.2.3) and mix gently. Adding the freezing medium drop by drop is crucial for cell recovery.
6. Pour 0.5 ml into a 1-ml cryogenic vial. A freezing ratio of 2×10^6 cells/vial is recommended.
7. Freeze overnight at –80°C in freezing boxes (Nalgene freezing box, C.N. 5100-0001).
8. Transfer to liquid nitrogen on the following day.

4.3.4 *Thawing hPSCs in Suspension*

The thawing procedure is identical to that used with feeders in adherent cultures.

1. Remove the vial from the liquid nitrogen.
2. Gently swirl the vial in a 37°C water bath.
3. When a small pellet of frozen cells remains, wash the vial in 70% ethanol.
4. Pipette the content of the vial up and down once to mix.
5. Place the contents of the vial in a conical tube and add, drop by drop, 2 ml of culture medium (see Sect. 4.2.1.1, 4.2.1.2 or 4.2.1.3). Adding the hESC medium drop by drop is crucial for cell recovery.
6. Centrifuge for 3 min at $80 \times g$ at a recommended temperature of 4°C.
7. Remove the supernatant and re-suspend the cells in 2 ml of culture medium (see Sect. 4.2.1.1, 4.2.1.2 or 4.2.1.3).
8. Place the cell suspension in a 35-mm petri dish and incubate at 37°C.

4.3.5 *Culturing hPSCs in a Dynamic System*

4.3.5.1 Culturing hPSCs in Erlenmeyer Flasks

1. Transfer cell spheres that were cultured in 1–3 petri dishes (10^6–6×10^6 cells) for at least one passage in suspension to 25 ml of culture medium in a 125-ml Erlenmeyer flask (Corning Incorporated, Corning, NY, USA) (see Sect. 4.2.1.1, 4.2.1.2 or 4.2.1.3). The optimal cell concentration is 10^6 cells/ml, but a lower cell concentration can also be employed.
2. Shake continuously at 60–90 rpm using an orbital shaker (S3.02.10L, ELMI Ltd, Riga, Latvia) and incubate at 37°C.

4.3.5.2 Culturing hPSCs in Spinner Flasks

1. Transfer the cell spheres cultured in 1–5 petri dishes (10^6–3×10^7 cells) for at least one passage in suspension to 30 ml of culture medium in a 100-ml glass ball spinner flask (CellSpin of Integra Biosciences, Fernwald, Germany) in (see

Sect. 4.2.1.1, 4.2.1.2 or 4.2.1.3). The optimal cell concentration is 10^6 cells/ml, but a lower cell concentration can also be employed.

2. Shake continuously at 60–90 rpm using a magnetic plate (CellSpin of Integra Biosciences, Fernwald, Germany) and incubate at 37°C. The morphology of spheres cultured in spinner flasks is illustrated in Fig. 4.5.

4.3.5.3 Splitting hPSCs Cultured in Dynamic Systems

1. Collect the cells cultured in suspension and place in a conical tube.
2. Centrifuge for 3 min at $80 \times g$ at a recommended temperature of 4°C.
3. Aspirate the medium from the tube without harming the cell pellet.
4. Re-suspend the cells in fresh medium (see Sect. 4.2.1.1, 4.2.1.2 or 4.2.1.3) and divide the cells equally in a new spinner or an Erlenmeyer flask. There is usually no need to dissociate the cells, but if so desired, passaging in petri dishes is preferable (see Sect. 4.3.2.2).
5. Add fresh medium (see Sect. 4.2.1.1, 4.2.1.2 or 4.2.1.3) and incubate at 37°C. A concentration of 1×10^6 cells/ml is recommended, but it is possible to start from 3×10^4 cells/ml and work up to 10×10^6 cells/ml.

4.3.6 Routine Culture of hPSCs in Suspension

When transferring PSCs from adherent to suspension culture, an adaptation time of 3–5 passages can be expected. During the adaptation time, some MEF clumps can still be observed (illustrated in Fig. 4.6), some of which may display disorganized morphology. In addition, during the adaptation time, background differentiation may exceed 5%. Examples of the morphology of differentiating spheres are illustrated in Fig. 4.7. When using petri dishes, the culture medium (see Sect. 4.2.1.1, 4.2.1.2 or 4.2.1.3) should be changed daily; if not feasible, add twice the amount of medium. The medium can be changed twice a week if a dynamic system is used, and if the cell concentration does not exceed 5×10^6 cells/ml. The cell concentration can be higher (up to 3×10^7 cells/ml), but then the medium should be changed on a daily basis. Our experience shows no differences in culture performance when using the different culture mediums presented (see Sect. 4.2.1.1, 4.2.1.2 or 4.2.1.3).

The medium can be changed as follows:

(a) If the cells are cultured using a 58-mm petri dish, they can be collected into a conical tube, centrifuged at $80 \times g$, and then re-plated using fresh culture medium. An alternative method is to tilt the dish gently to centralize the cells, aspirate as much medium as possible without harming the cells, and add 5 ml of fresh medium.

(b) If the cells are cultured in Erlenmeyer or spinner flasks, remove the dish from the magnetic plate, let the cells sink, then aspirate as much medium as possible, and replace with fresh medium. Alternatively, the medium can be changed by centrifugation.

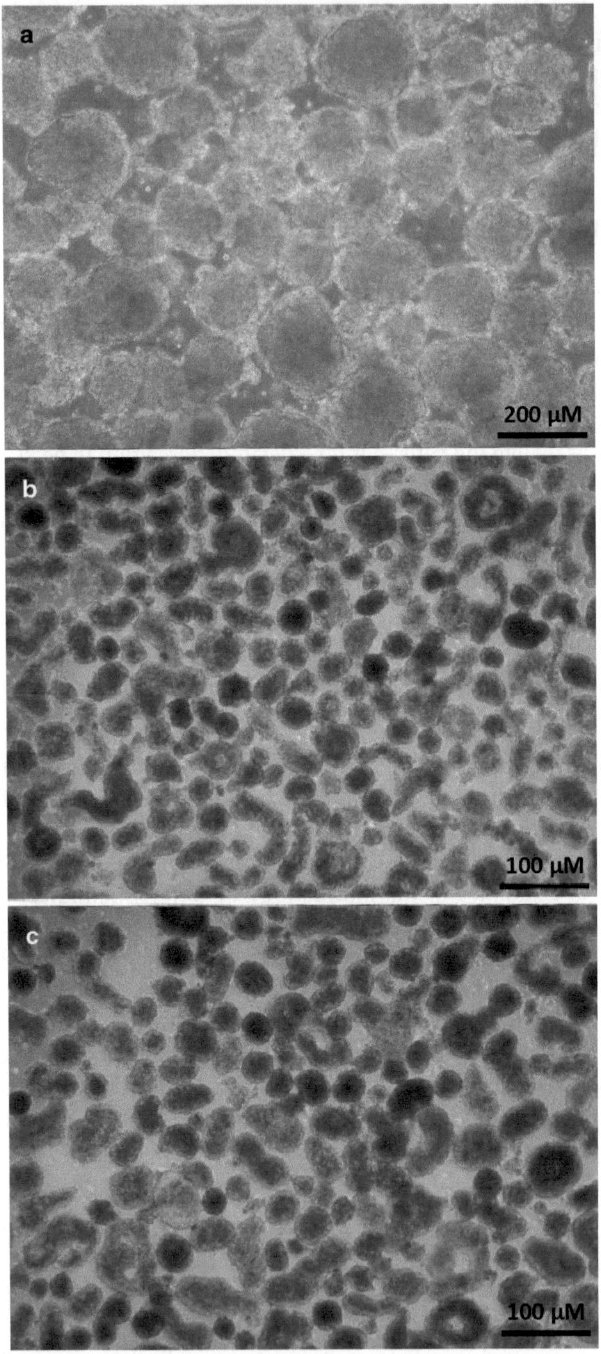

Fig. 4.5 Dynamic culture of hPSCs in spinner flasks. The cells spontaneously form homologous spheres with a diameter range of 100–300 μM. (**a**) hESC line H14 cultured in spinner flasks for 2 weeks. (**b**, **c**) hESC line I3 cultured in spinner flasks for 3 months. (**a**) Bar 200 μM and (**b**, **c**) bar 100 μM

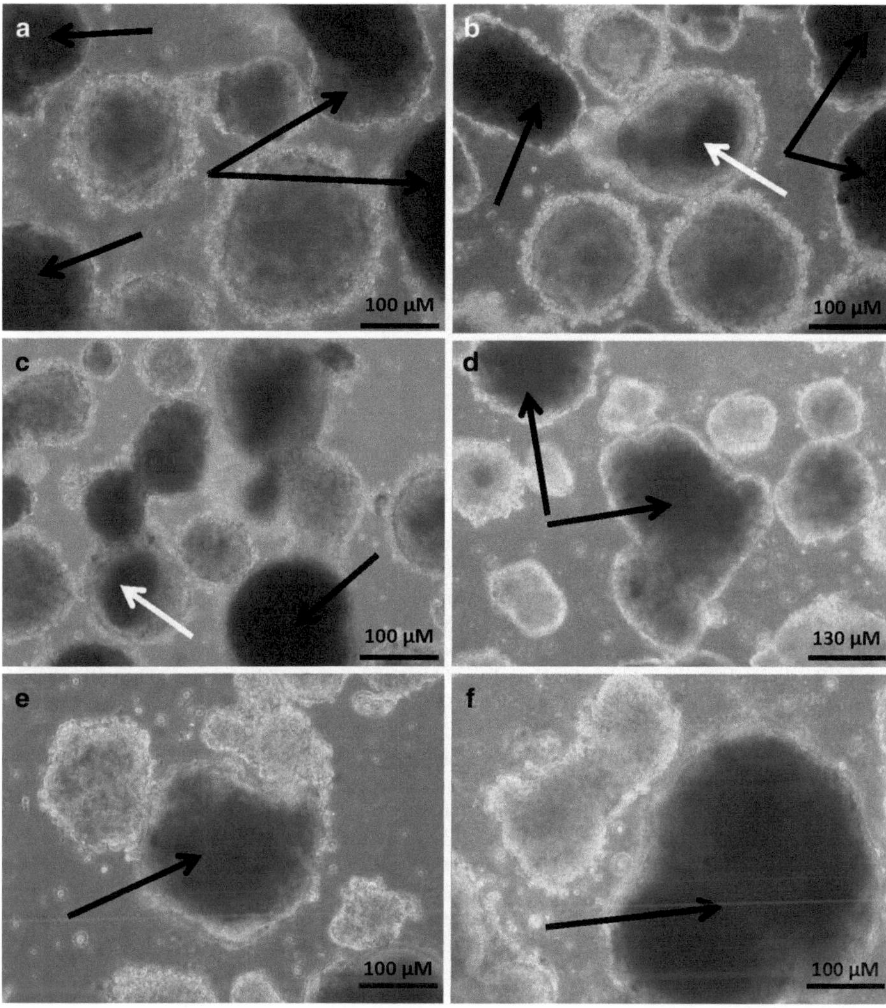

Fig. 4.6 Morphology of spheres during adaptation to suspension culture. (**a–c**) hESC line H14 cultured in suspension for 4 passages. Some of the clumps in the culture have *brown* color (marked with *black arrows*) and consist of either dead cells or MEFs. One spheroid has an apoptotic center (marked with *white arrow*). (**d–f**) hiPSC line iLBWT30m (derived from skin fibroblast) cultured in suspension for 8 passages. Several dark clumps (marked with *black arrows*) and some clumps with apoptotic center (marked with *white arrow*) still exist. It seems that this line requires more time to adapt to the suspension culture than does the average line. (**a–c**) Bar 100 μM, (**d**) bar 130 μM, and (**e**, **f**) bar 100 μM

As with cells cultivated in adherent culture, the hPSCs should be passaged (see Sect. 4.3.2) every 5–8 days. Examples of sphere morphology, post-mechanical splitting, and trypsin dissociation are demonstrated in Figs. 4.3 and 4.4, respectively. If the culture is not split, in addition to an increase in differentiation rate, the spheres

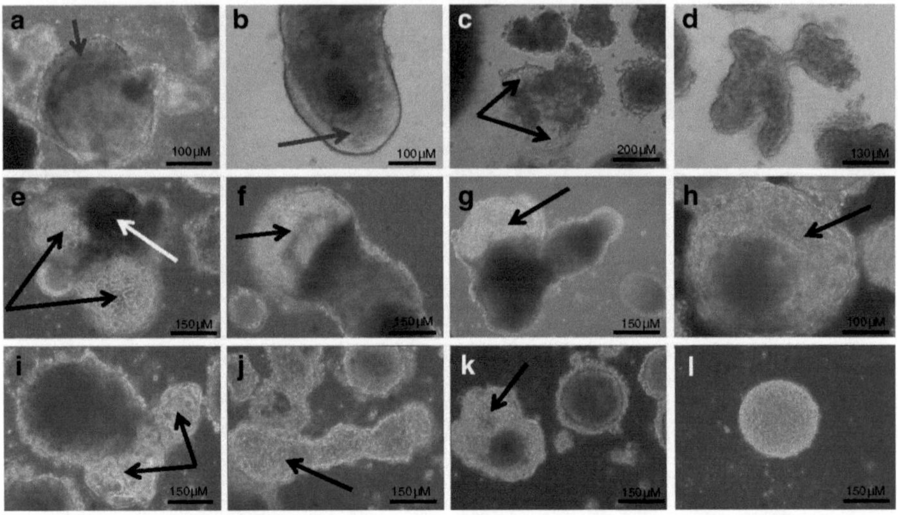

Fig. 4.7 Differentiated spheres cultured in suspension. (**a**) hiPSC line mKTR13.4 cultured in suspension for 5 passages. The cells formed epithelia at the clump borders (marked with *gray arrow*). (**b**) hiPSC line iLBWT30m cultured in suspension for 8 passages. The cells formed neuron-like epithelia at the edges of the clump (marked with *gray arrow*). (**c**) hiPSC line mKTR13.4 cultured in suspension for 5 passages. The cells formed a cyst (marked with *black arrow*). (**d**) hiPSC line hCB1 derived from cord blood cells cultured in suspension for 34 passages. The cells formed neuron-like epithelia. (**e**) hESC line I3.2 cultured in suspension for 5 passages. A cyst structure was formed (marked with *black arrow*). The clump also contains pigment cells (marked with *white arrow*). (**f, g**) hESC line H7 cultured in suspension for 8 passages. The cells formed a cyst (marked with *black arrow*). (**h**) hESC line I3 cultured in suspension for 1 passage. The cells formed a cyst (marked with *black arrow*). (**i**) hPSC line J6 cultured in suspension for 5 passages. The cells formed a cyst (marked with *black arrow*). (**j**) hiPSC line mKTR13.4 cultured in suspension for 5 passages. The cells formed an elongated cyst (marked with *black arrow*). (**k**) hESC line I3 cultured in suspension for 1 passage. The cells formed a cyst (marked with *black arrow*). (**l**) hiPSC line iLBWT30m cultured in suspension for 8 passages. The cells formed a compacted neurosphere, with a ball-like structure instead of a disc-like one. (**a, b**) and (**h**) Bar 100 μM, (**c**) bar 200 μM, (**d**) bar 130 μM, and (**e–g**) and (**i-l**) bar 150 μM

may grow to a size of 300 μM or more, which would interfere with diffusion and lead to apoptosis in the center of the sphere (demonstrated in Fig. 4.8).

Cell features can be tested with the adherent culture assays, including fluorescence-activated cell sorting (FACS) analysis for pluripotency cell surface markers, immunostaining for undifferentiation markers (examples provided in Chap. 9), and gene expression assays. Similar to adherent cells, the culture can be evaluated either by direct monitoring under a microscope (examples for undifferentiated and differentiated morphology are illustrated in Figs. 4.1 and 4.7, respectively), or by re-plating part of the spheres on MEFs. In the latter case, colonies with undifferentiated morphology will emerge within 48 h post-plating (shown in Figs. 4.9 and 4.10).

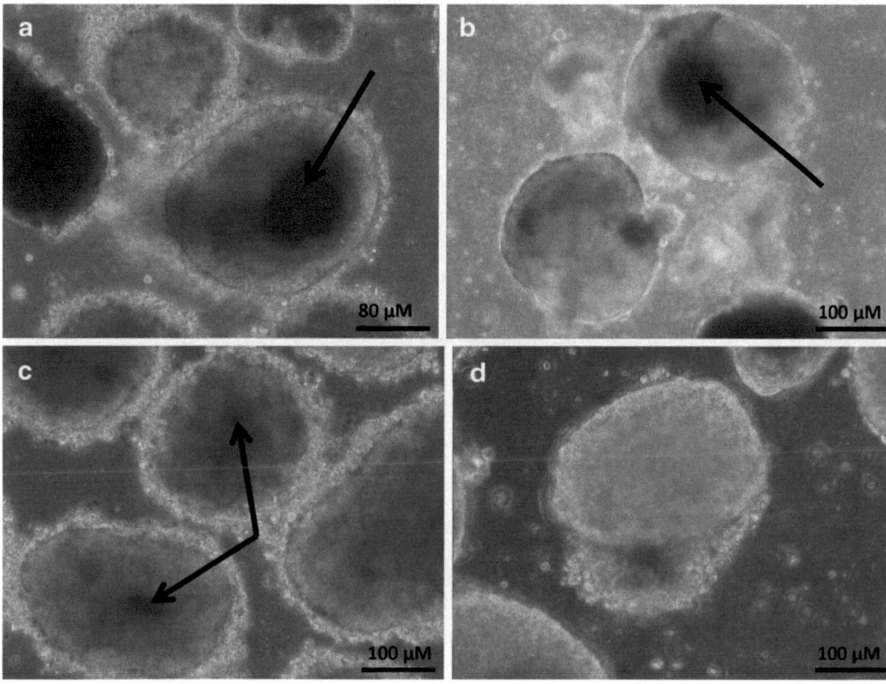

Fig. 4.8 Spheres containing an apoptotic center. In some cases, when spheres reach a certain size, diffusion is insufficient and an area of apoptotic cells appears at their center. (**a**) hESC line H14 cultured in suspension for 4 passages. The apoptotic center is marked with a black arrow. (**b**) hiPSC line mKTR13.4 cultured in suspension for 5 passages. The apoptotic center is marked with a black arrow. (**c**) hESC line I3 cultured in suspension for 43 passages. A very small apoptotic center was formed (marked with *black arrow*). (**d**) hiPSC line iLBWT30m cultured in suspension for 8 passages without an apoptotic center. (**a**) Bar 80 μM, and (**b**–**d**) bar 100 μM

Additionally, the sphere morphology can be examined by histological slides (demonstrated in Fig. 4.11).

If transferred, the cells can be sent frozen or fresh. Fresh cells should be sent using 50-ml tubes filled with 25 ml culture medium. Examples of sphere morphology post-shipment are illustrated in Fig. 4.12.

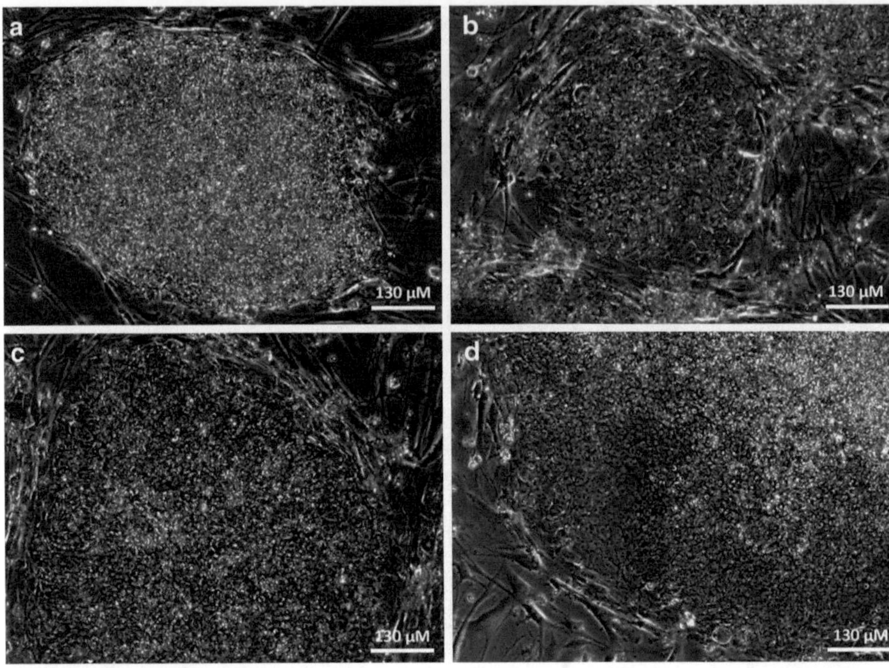

Fig. 4.9 Colonies formed by spheres re-cultured on MEFs. When spheres cultured in suspension are re-cultured on MEFs, they form colonies with typical hESC morphology, i.e., round, with clear borders and containing small cells with large nuclei. This occurs within 48 h of re-culturing. (**a**) A colony formed by hESC line I4 cultured in suspension for 10 passages after re-plating on MEFs. (**b**, **c**) A colony formed by hESC line I3 cultured in suspension for 11 passages after re-plating on MEFs. (**d**) A colony formed by hESC line I6 cultured in suspension for 10 passages after re-plating on MEFs. Bar 130 μM

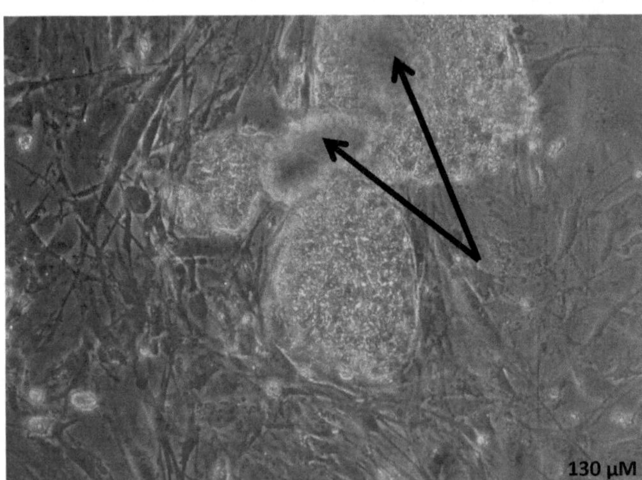

Fig. 4.10 Colonies formed by spheres re-cultured on MEFs after over 2 years of culture in suspension. Colonies formed by hESC line I3 cultured in suspension for 121 (!) passages after re-plating on MEFs. Some spheres failed to attach to the MEFs (marked with *black arrows*). Bar 130 μM

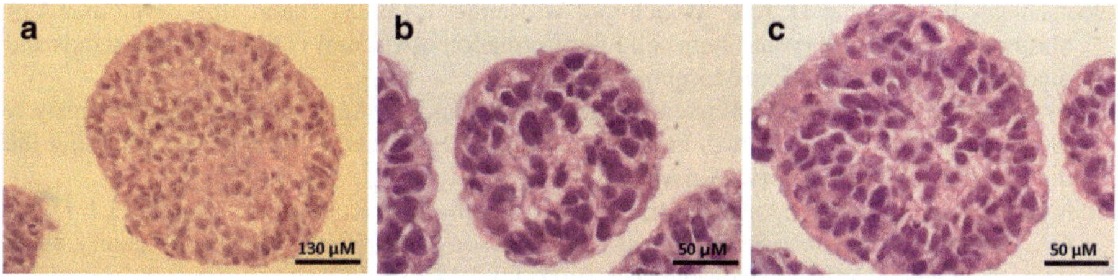

Fig. 4.11 Spheroid morphology in histological sections. (**a**) Spheres created by hESC line I4 after 48 passages in suspension. (**b**, **c**) A histological section of spheres formed by I6 cells cultured for 30 passages in suspension. The spheres contained relatively homogeneous cell populations of small round cells with large nuclei, typical of pluripotent stem cells. Differentiating cells could not be observed. (**a**) Bar 130 μm and (**b**, **c**) bar 50 μm

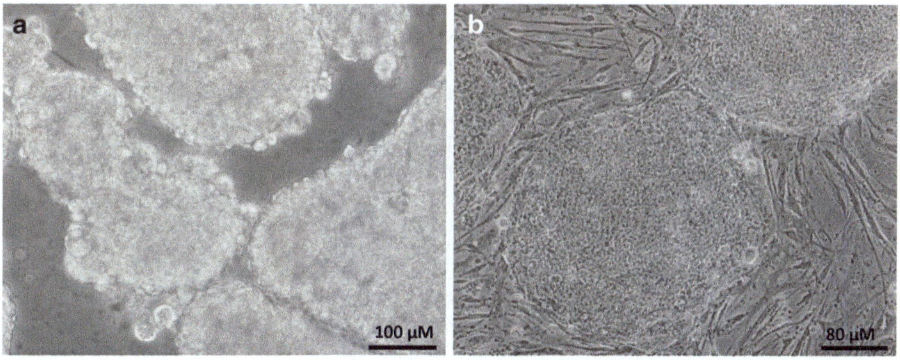

Fig. 4.12 Recovery of fresh cells after shipment in suspension. (**a**) Morphology of the spheres of hESCs I3 shipped as living cells. The spheres have undifferentiated morphology with no sign of apoptosis. (**b**) Colonies formed after the same spheres were re-cultured with MEFs. The colonies contain undifferentiated cells. (**a**) Bar 100 μM and (**b**) bar 80 μM

References

Amit M, Carpenter MK, Inokuma MS, Chiu CP, Harris CP, Waknitz MA, Itskovitz-Eldor J, Thomson JA (2000) Clonally derived human embryonic stem cell lines maintain pluripotency and proliferative potential for prolonged periods of culture. Dev Biol 227:271–278

Amit M, Margulets V, Segev H, Shariki C, Laevsky I, Coleman R, Itskovitz-Eldor J (2003) Human feeder layers for human embryonic stem cells. Biol Reprod 68:2150–2156

Amit M, Shariki K, Margulets V, Itskovitz-Eldor J (2004) Feeder and serum-free culture system for human embryonic stem cells. Biol Reprod 70:837–845

Amit M, Chebath J, Marguletz V, Laevsky I, Miropolsky Y, Shariki K, Peri M, Revel M, Itskovitz-Eldor J (2010) Suspension culture of undifferentiated human embryonic and induced pluripotent stem cells. Stem Cell Rev Rep 6:248–259

Amit M, Laevsky I, Miropolsky Y, Shariki K, Peri M, Itskovitz-Eldor J (2011) Dynamic suspension culture for scalable expansion of undifferentiated human pluripotent stem cells. Nat Protoc 6:572–579

Doetschman TC, Eistetter H, Katz M, Schmidt W, Kemler R (1985) The *in vitro* development of blastocyst-derived embryonic stem cell lines: formation of visceral yolk sac, blood islands and myocardium. J Embryol Exp Morphol 87:27–45

Itskovitz-Eldor J, Schuldiner M, Karsenti D, Eden A, Yanuka O, Amit M, Soreq H, Benvenisty N (2000) Differentiation of human embryonic stem cells into embryoid bodies comprising the three embryonic germ layers. Mol Med 6:88–95

Ludwig TE, Levenstein ME, Jones JM, Berggren WT, Mitchen ER, Frane JL, Crandall LJ, Daigh CA, Conard KR, Piekarczyk MS, Llanas RA, Thomson JA (2006) Derivation of human embryonic stem cells in defined conditions. Nat Biotechnol 24:185–187

Richards M, Fong CY, Chan WK, Wong PC, Bongso A (2002) Human feeders support prolonged undifferentiated growth of human inner cell masses and embryonic stem cells. Nat Biotechnol 20:933–936

Oh SK, Chen AK, Mok Y, Chen X, Lim UM, Chin A, Choo AB, Reuveny S (2009) Long-term microcarrier suspension cultures of human embryonic stem cells. Stem Cells Res 2:219–230

Olmer R, et al. (2010) Long term expansion of undifferentiated human iPS and ES cells in suspension culture using a defined medium. Stem Cell Res 5:51–64

Phillips BW, Horne R, Lay TS, Rust WL, Teck TT, Crook JM (2008) Attachment and growth of human embryonic stem cells on microcarriers. J Biotechnol 138:24–32

Singh H, Mok P, Balakrishnan T, Rahmat SN, Zweigerdt R (2010) Up-scaling single cell-inoculated suspension culture of human embryonic stem cells. Stem Cell Res 4:165–179

Steiner D, et al. (2010) Derivation, propagation and controlled differentiation of human embryonic stem cells in suspension. Nat Biotech 28:361–364

Takahashi K, Tanabe K, Ohnuki M, Narita M, Ichisaka T, Tomoda K, Yamanaka S (2007) Induction of pluripotent stem cells from adult human fibroblasts by defined factors. Cell 131:861–872

Yu J, Vodyanik MA, Smuga-Otto K, Antosiewicz-Bourget J, Frane JL, Tian S, Nie J, Jonsdottir GA, Ruotti V, Stewart R, Slukvin II, Thomson JA (2007) Induced pluripotent stem cell lines derived from human somatic cells. Science 318:1917–1920

Zandstra PW, Nagy A (2001) Stem cell bioengineering. Annu Rev Biomed Eng 3:275–305; review

Chapter 5
Differentiation of Pluripotent Stem Cells In Vitro: Embryoid Bodies

Abstract Human pluripotent stem cells (hPSCs), both embryonic and induced, are traditionally cultured as adherent cells. When transferred to suspension culture, they normally form cell aggregates known as embryoid bodies (EBs). As implied by their name, EBs consist of differentiating cells representing the three embryonic germ layers. The cascade of events that transpire within these cells mimic early embryonic development. Therefore, with its crucial role in increasing differentiation efficiency and velocity, the stage of EB formation is considered an important step in many directed and spontaneous differentiation protocols. This chapter focuses on the morphology of EBs after their formation in static and dynamic cultures.

5.1 Introduction

Human pluripotent stem cells (hPSCs), both embryonic and induced, are cultured as adherent cells in conditions supporting undifferentiated growth (Thomson et al. 1998; Takahashi and Yamanaka 2006). When hPSCs are transferred from an environment supporting self-renewal to tissue culture-treated dishes (low adherent plates) comprising suspension culture, they spontaneously form cell aggregates known as embryoid bodies (EBs). EBs demonstrate regional differentiation into embryonically distinct cell types (Doetschman et al. 1985; Itskovitz-Eldor et al. 2000; Takahashi and Yamanaka 2006). Small aggregates can be observed 24 h after plates are in suspension. In mouse embryonic stem cells (ESCs), these aggregates tend to organize into a special structure consisting of an internal endoderm layer and an external ectoderm layer; however, hPSCs tend to be somewhat less organized (Itskovitz-Eldor et al. 2000). Four to ten days after being placed in suspension, some of the EBs become cavitated, and by 2 weeks some become cystic

M. Amit and J. Itskovitz-Eldor, *Atlas of Human Pluripotent Stem Cells: Derivation and Culturing*, Stem Cell Biology and Regenerative Medicine, DOI 10.1007/978-1-61779-548-0_5, © Springer Science+Business Media, LLC 2012

(Itskovitz-Eldor et al. 2000). Similar to mouse EBs, EBs formed by hPSCs replicate aspects of early development, displaying differentiation into derivatives of all three embryonic germ layers. Though complex tissue structures are rarely seen (Itskovitz-Eldor et al. 2000), the EB stage has become a major step in inducing hPSC differentiation.

Many differentiation protocols for hPSCs rely on EB formation. For example, the re-plating of spontaneously differentiating EBs on an adherent substrate containing beating areas generates cells exhibiting structural and functional properties of early stage cardiomyocytes (Kehat et al. 2001, 2002; Germanguz et al. 2011; Zhang et al. 2009; Tanaka et al. 2009; Zwi et al. 2009). Differentiation into cardiomyocytes can also be induced. One of the first studies demonstrating cardiomyocyte enrichment was based on plating 4–6-day-old EBs, treating them with 5-aza-2′-deoxycytidine, and separating the resultant cardiac cells by dissociation and by Percoll gradient density centrifugation. The result was a population of 70% pure cardiomyocytes (Xu et al. 2002a). Additional factors, such as Activin A and BMP4, were found to promote cardiac differentiation of hPSCs. A number of protocols that include the EB stage are based on supplementation of these factors (Laflamme et al. 2007). More complex protocols, based on the step-by-step provision of certain cultural conditions, have recently emerged, thus leading the generation of cardiomyocyte progenitors (Yang et al. 2008).

Spontaneously formed EBs derived from hESCs have been shown to contain endothelial cells (Levenberg et al. 2002; Gerecht-Nir et al. 2003) and low levels of CD45-positive cells. Most co-express CD34, a phenotype similar to the first definitive hematopoietic cell. CD34 can be detected within the wall of the dorsal aorta of developing human embryos (Chadwick et al. 2003). A team of investigators showed that treating hEBs with hematopoietic cytokines induces differentiation of hematopoietic cells (Chadwick et al. 2003), whereas treating them with vascular endothelial growth factor A (VEGF-A_{165}) selectively induces erythropoietic differentiation (Cerdan et al. 2004). In addition, spontaneously formed EBs were shown to contain neural progeny such as neurons, astrocytes, and oligodendrocytes (Zhang et al. 2001). More recent protocols include the induction of hPSC neural lineage differentiation by the use of specific factors, such as proliferating neural progenitors, which are capable of differentiation into neurons (both GABAergic and dopaminergic), astrocytes, and oligodendrocytes (Koch et al. 2009; Gerrard et al. 2005). EBs formed by hPSCs were also shown to contain extra embryonic endoderm, including trophoblast cells, which can be induced by plating EBs on Matrigel (Gerami-Naini et al. 2004).

However, hPSCs can also be induced to differentiate without the EB stage. When hPSCs are removed from culture conditions that promote their self-renewal and are plated as crowded colonies on gelatin, they spontaneously differentiate. The same happens when hPSCs are grown to confluency on mouse embryonic fibroblasts (MEFs). Therefore, to direct the differentiation of hPSCs to a specific lineage, two-dimensional (2D) culture conditions should be modified. Indeed, directed differentiation methods for hPSCs have been established, enabling researchers to obtain enriched or purified specific cell populations, such as cardiac cells, neurons, endothelial cells, and trophoblasts.

Cardiac-directed differentiation in 2D culture, without an EB stage, can be achieved by the co-culture of hPSCs with visceral endoderm-like cells. This results in the contraction of areas containing cells characterized by cardiac features (Mummery et al. 2002, 2003). Other efficient protocols to direct cardiomyocyte-like cell differentiation of hPSCs are based on culturing the cells with Matrigel matrix and exposing them on specific days to BMP4 and Activin A (Laflamme et al. 2007). Endothelium-like cells can also be obtained from hPSC differentiation by culturing the stem cells with collagen type IV (Gerecht-Nir et al. 2003). Similarly, trophoblast differentiation was induced by a feeder layer-free culture of hPSCs with the supplementation of BMP-4 (Xu et al. 2002b). Two-dimensional cultures have also been used for neuronal progenitor derivation by means of co-culture with suitable feeder cells that are capable of differentiating into different neural lineages (Perrier et al. 2004; Buytaert-Hoefen et al. 2004). Moreover, co-cultures with specific feeder layers have been used to derive hematopoietic cells from hPSCs (Kaufman et al. 2001; Vodyanik et al. 2005).

For clinical and industrial applications of hPSCs, mass production of the desired cells is necessary. Scalable cultures of both undifferentiated and differentiated cells are needed, preferably by using controlled bioreactors. To this end, three possible protocols for promoting hPSC differentiation within EBs have been developed (Gerecht-Nir et al. 2004; Dang et al. 2004; Cameron et al. 2006; Yirme et al. 2008). For one, Gerecht-Nir et al. achieved dynamic formation of EBs in rotating bioreactors containing representative tissues of the three embryonic germ layers (Gerecht-Nir et al. 2004; Cameron et al. 2006). Another method for controlling EB formation involves mass encapsulation of hPSCs using agarose to prevent agglomeration and to control EB size, followed by culturing of the resultant EBs in spinner flasks or reactors (Dang et al. 2004). Yirme et al. developed a simple and straightforward method to culture EBs in spinner flasks. However, the fact that the initiating step is in a static culture limits the scalability of the system [Yirme et al, 2008]. Further investigation of means to regulate EB formation is needed.

5.2 Materials for EB Formation

5.2.1 Culture Medium Supplemented with Serum

80% DMEM/F12 (Biological Industries Ltd), 20% fetal bovine serum (FBS, HyClone C.N. SV30160-03), 1% nonessential amino acid (Invitrogen Corporation), 2 mM L-glutamine (Invitrogen Corporation), and 0.1 mM β-mercaptoethanol (Invitrogen Corporation).

5.2.2 Splitting Medium Based on Collagenase

The splitting medium consists of 1 mg/ml collagenase type IV (Wordington, type IV C.N 4189, activity of 220–320 unite/mg) in DMEM (any type can be used).

5.3 Methods for EB Formation and Culture

5.3.1 EB Formation

The plating of high-density undifferentiated cells is recommended for initiating EB formation (5×10^6 to 10^7 at 60 cm^2).

1. Aspirate the medium from the wells that are to be split. Add splitting medium (see Sect. 5.2.2) to cover the wells (0.5 ml for 10 cm^2), and incubate for 20–40 min. Most colonies will float.
2. Add 1 ml of culture medium (see Sect. 5.2.1) and carefully collect the cells.
3. Collect the cell suspension and place in a conical tube.
4. Centrifuge for 3 min at $80 \times g$ at a recommended temperature of 4°C.
5. Re-suspend cells in media, break the cells by moving a 200-μl tip up and down 5–10 times, and plate in 58-mm petri dishes.

5.3.2 Routine Culture of EBs

Initially, the culture medium (see Sect. 5.2.1) should be changed twice weekly. As the cell concentration increases, the changes should be more frequent, until they are daily. Alternatively, the EBs can be split into two petri dishes. Twenty-four hours post-EB formation, small cell aggregates appear, which may organize into a special structure consisting of an internal endoderm layer and an external ectoderm layer (Fig. 5.1). Three days later, the aggregates increase in size but the morphology does not change, except that the particular organization described above disappears (Fig. 5.2). About 4 days after EB formation, some of the EBs that have formed develop a cavity (Fig. 5.3). Fourteen days post-formation, cystic EBs are visible (Fig. 5.3), and lineage differentiation is easily demonstrated by staining or by gene expression assays. EBs older than 14 days may contain more complex structures, some of which can be identified by morphological changes (Fig. 5.4), and others in histological sections (Figs. 5.5 and 5.6). In these EBs, a variety of cell types can be noted (Fig. 5.7). For some induced differentiation protocols, EBs can be re-plated on coated plates (with gelatin, fibronectin, or Matrigel matrix, according to the differentiation protocol). The EBs that are attached and also cell outgrowth become visible (Fig. 5.8). In some of the EBs, cysts fade, and in some, they re-form as

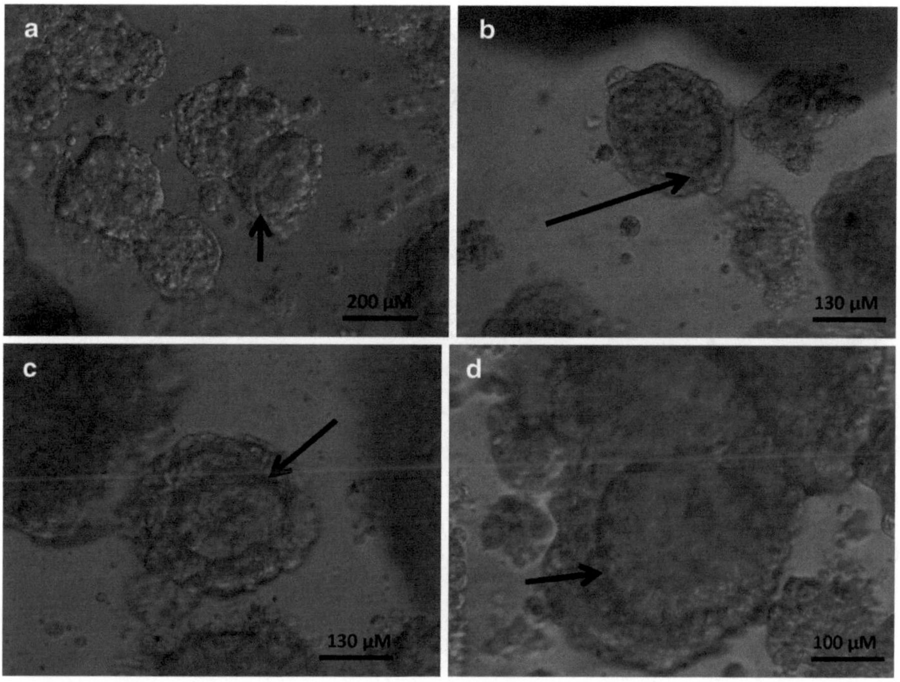

Fig. 5.1 Formation of an internal ectoderm layer and an outer endoderm layer. Embryoid bodies (EBs) formed by hESC line CL1, 48 h post-formation. Small cell aggregates can be seen, some consisting of an internal endoderm layer and an external ectoderm layer (marked with *black arrow*). (**a**) Bar 200 μM, (**b, c**) bar 130 μM, and (**d**) bar 100 μM

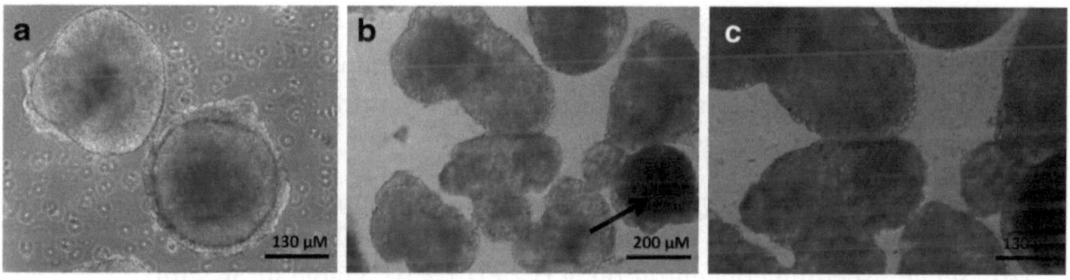

Fig. 5.2 Three-day-old EBs. (**a**) Formed by BG01 hESC line; note that the external endoderm layer is fading. (**b**) Formed by hESC CL1 line; note that most EBs seem to contain homological cell populations with no signs of clear differentiation and cavity formation. An aggregate with necrotic cells is marked (*black arrow*). (**c**) Higher magnification of (**b**). (**a–c**) Bar 130 μM, (**b**) bar 200

adherent cells (Figs. 5.8 and 5.9). Figure 5.10 shows the cell types resulting from plated EBs, including neuron-like cells, rosette structure, and lipid containing cells. In some cases, EBs cultured in suspension tend to attach to each other. Figure 5.11 shows a histological section of attached EBs, demonstrating cell transfer between the two EBs.

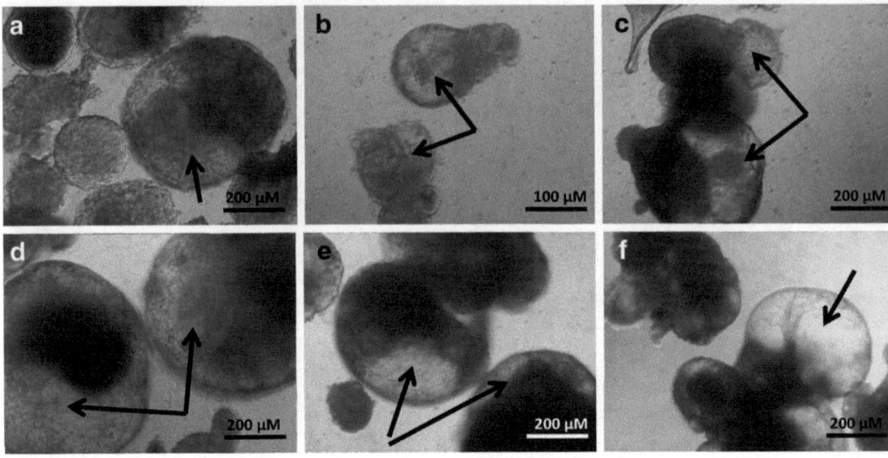

Fig. 5.3 Cavity formation by EBs. EBs formed by hESC line CL1. (**a, b**) Cavity formation by 4-day-old EBs (marked with *black arrows*). (**c, f**) Cystic EBs 10 and more days post-formation. Cysts are marked with *black arrows*. (**a, c–f**) Bar 200 μM, (**b**) bar 100 μM

5.3.3 Culturing EBs in Spinner Flasks

The protocol proposed is based on that developed by Yirme and colleagues (Yirme et al. 2008). Although this method still requires the formation of EBs in a static culture, it saves time and effort during culturing.

1. Follow steps for EB formation, as described in Sect. 5.3.1. A recommended clump size for initiating EBs is 1,445 ± 115 cells/clump.
2. At 24–48 h post-EB derivation, transfer the resulting EBs to a spinner flask according to the following parameters:

    (a) Use a scalable and reliable stirred tank bioreactor (STR) method with mild shear force (i.e., GBI impeller type, Erlenmeyer).
    (b) Set cell seeding concentration at $0.7 \pm 0.1 \times 10^6$ viable hPSCs/ml.
    (c) Propagate in a stirred system at a speed of 70 rpm, within an incubator at 37° and 5% CO_2.
    (d) Set linear volumetric scaling in the GBI spinner flask to the highest manufacture volume recommended.
    (e) Change the medium twice weekly.

The resulting EBs seem more homogeneous in size (Fig. 5.12a) and contain complex structures and a variety of cell types, similar to the EBs formed and cultured in 2D. The morphology of the resulting EBs is exemplified in Fig. 5.12.

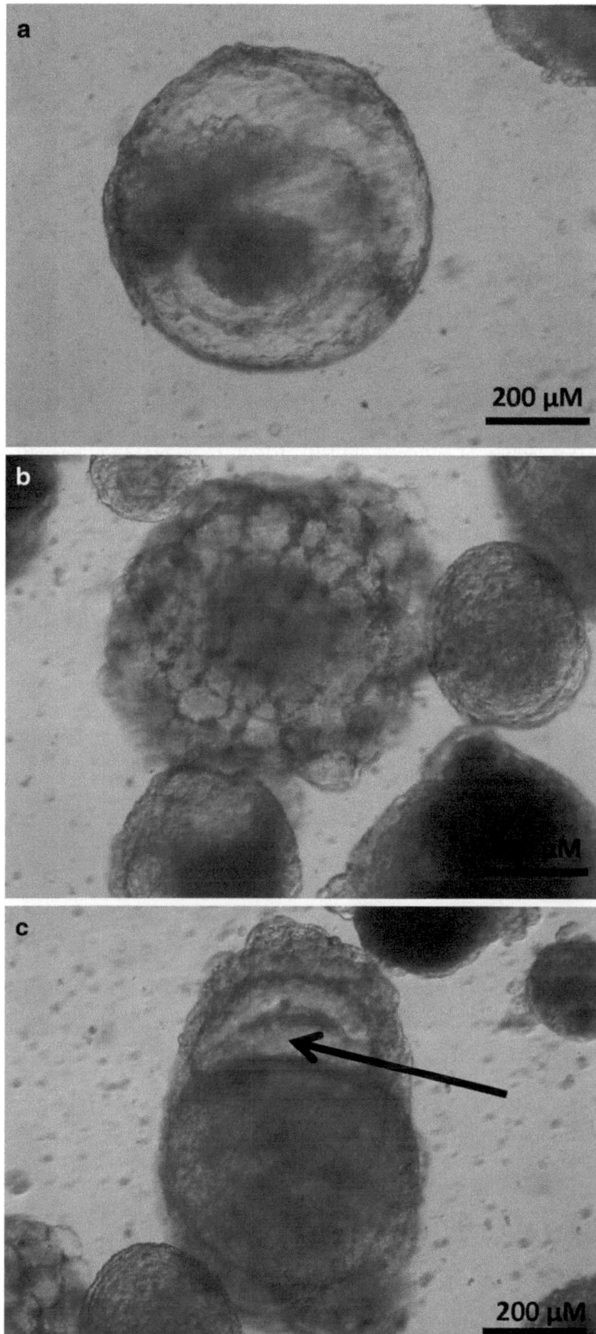

Fig. 5.4 Complex structures formed within EBs. 14-day-old and above EBs formed by hESC line CL1, after being cultured in suspension as undifferentiated cells. (**a**) With more than one cyst, (**b**) with ring-like structures probably created by formed capillary, and (**c**) containing unique epithelial structure (marked with *black arrow*). Bar 200 μM

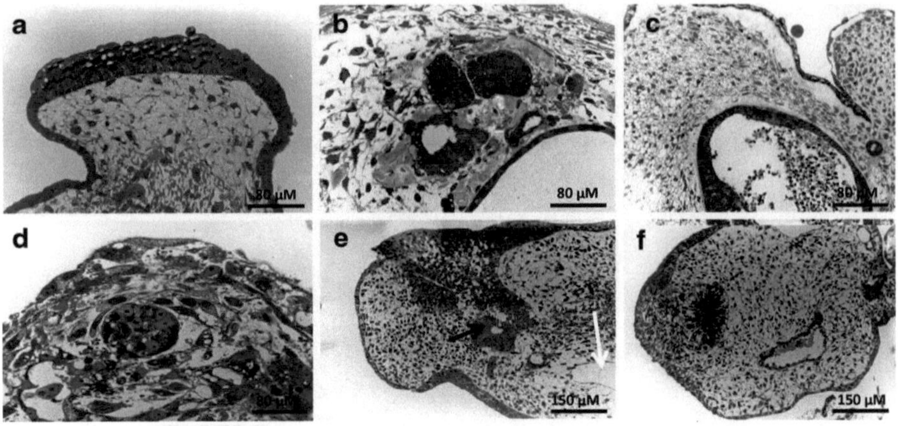

Fig. 5.5 Histological section of cultured old EBs. EBs formed by hESC line H9.2, cultured for at least 1 month. (**a**) Stratified epithelium at the periphery of EB, and mesenchyma-like cells. (**b**) Four balls of differentiation, two with a small lumen, and a cube epithelium facing a large lumen. (**c**) Stratified epithelium lining a lumen in which groups of undefined cells can be seen. (**d**) A ball of differentiating epithelium-like cells. (**e**) EB surrounded by stratified epithelium, containing mesenchyma-like cells, a tube-like structure lined by stratified epithelium (*black arrow*), and a primitive blood vessel (*white arrow*). (**f**) Stratified epithelium at the periphery of EB, and a ball-like structure of undefined differentiating cells. (**a–d**) Bar 80 μM, (**e, f**) bar 150 μM. 1-μM Epon histological sections stained with alkaline toluidine blue

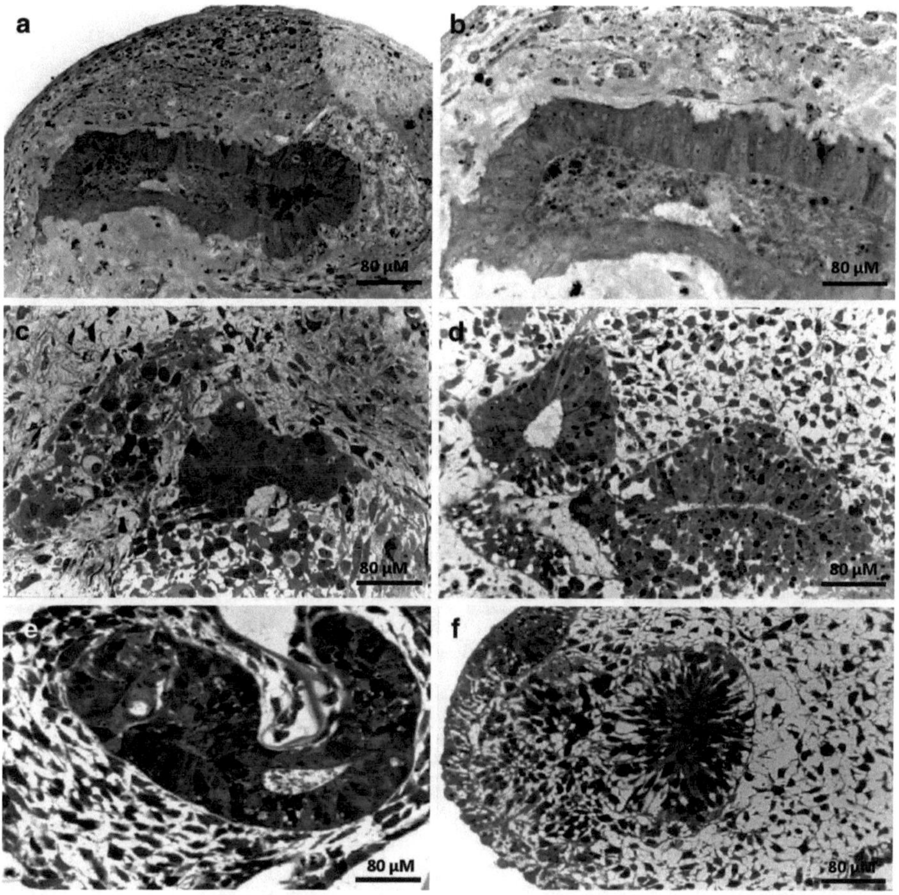

Fig. 5.6 Complex structures in a histological section of cultured EBs. EBs formed by hESC line H9.2, cultured for at least 1 month. (**a**) Columnar epithelium lining a lumen. (**b**) Same as (**a**), at higher magnification. (**c**) Aggregate of undefined differentiating cells surrounded by mesenchyma-like cells. (**d**) Tube-like structures lined by stratified epithelium. (**e**) Aggregate of undefined differentiating cells lining a small lumen. (**f**) Stratified epithelium at the periphery of EB, also containing an aggregate of undefined differentiating cells. Bar 80 μM. 1-μM Epon histological sections stained with alkaline toluidine blue

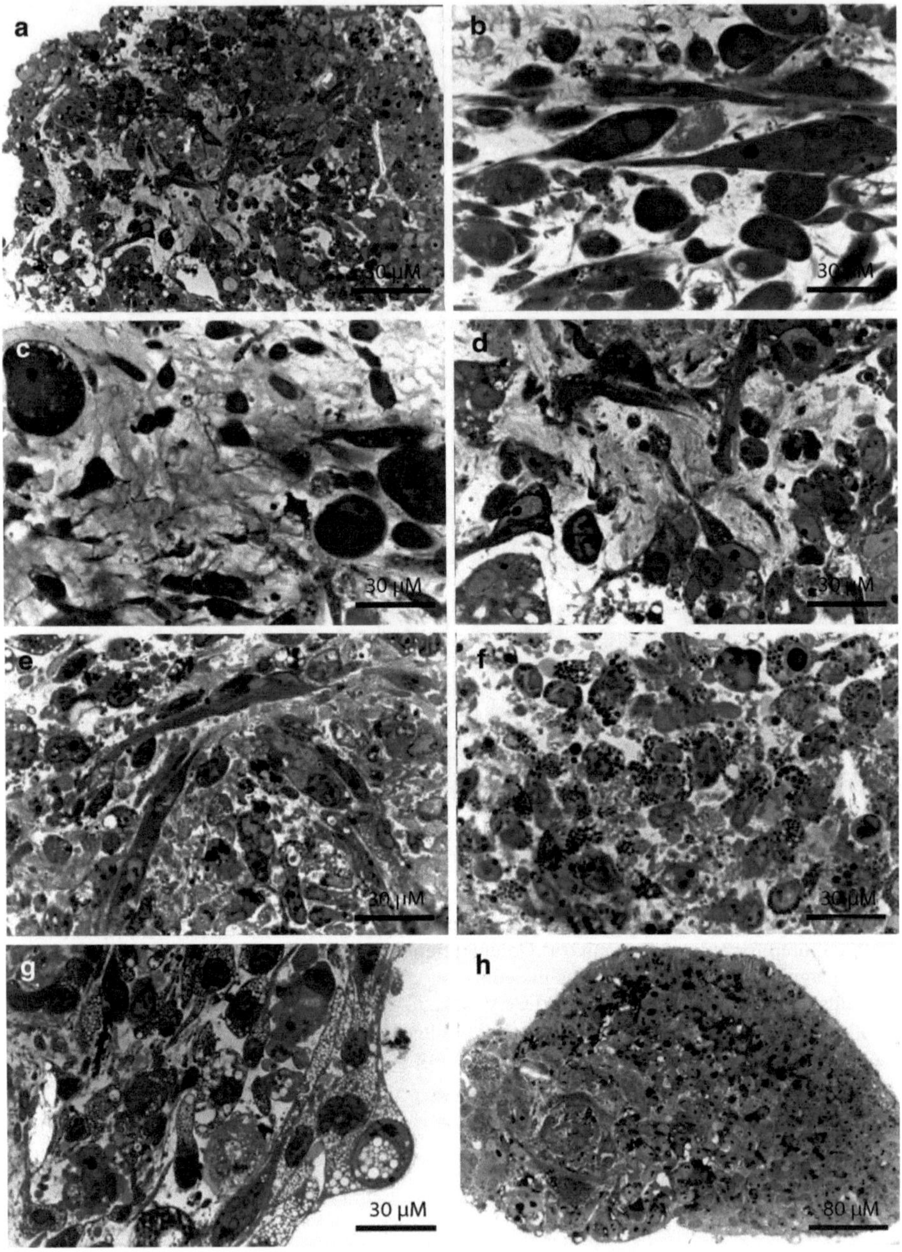

Fig. 5.7 Cell types in old EBs. EBs formed by hESC line H9.2, cultured for at least 1 month. (**a**, **c**) Undefined differentiating cells within EBs, some with elongated structure (**b**, **d**, **e**), some containing dark vesicles (**f**), and some containing lipid drops (**g**). In some cells, a large regular nucleus with pronounced nucleoli can be observed, which indicates intense synthetic activities (**h**). (**a**, **h**) Bar 80 μM, (**b–g**) bar 30 μM. 1-μM Epon histological sections stained with alkaline toluidine blue

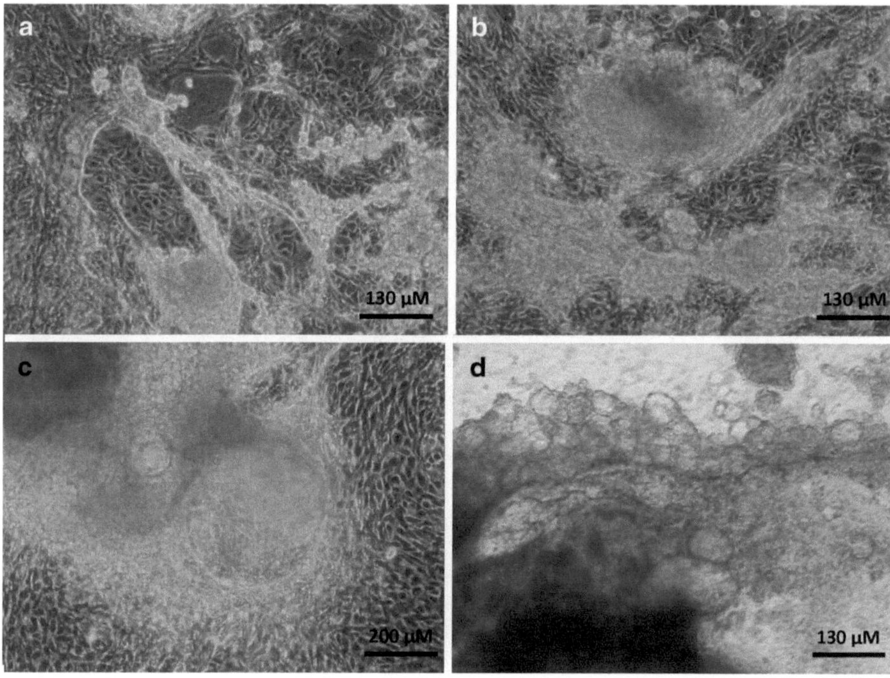

Fig. 5.8 Plated EBs. 10-day-old EBs were re-plated on gelatin. (**a**, **b**) Outgrowth from EBs formed by hESCs I3.2. (**c**, **d**) Plated EBs created by hESC line CL1 containing cysts. (**a**, **b**, **d**) Bar 130 μM, (**c**) bar 200 μM

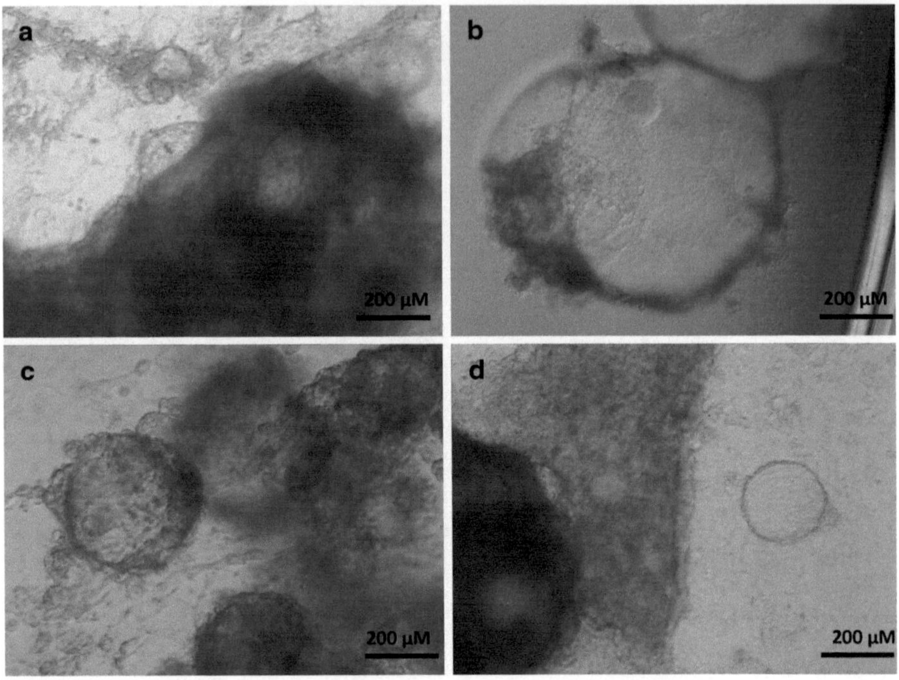

Fig. 5.9 Plated EBs containing cysts. 10-day-old and above EBs formed by hESC line CL1, after being cultured in suspension as undifferentiated cells, were re-plated on gelatin-coated plates. Bar 200 μM

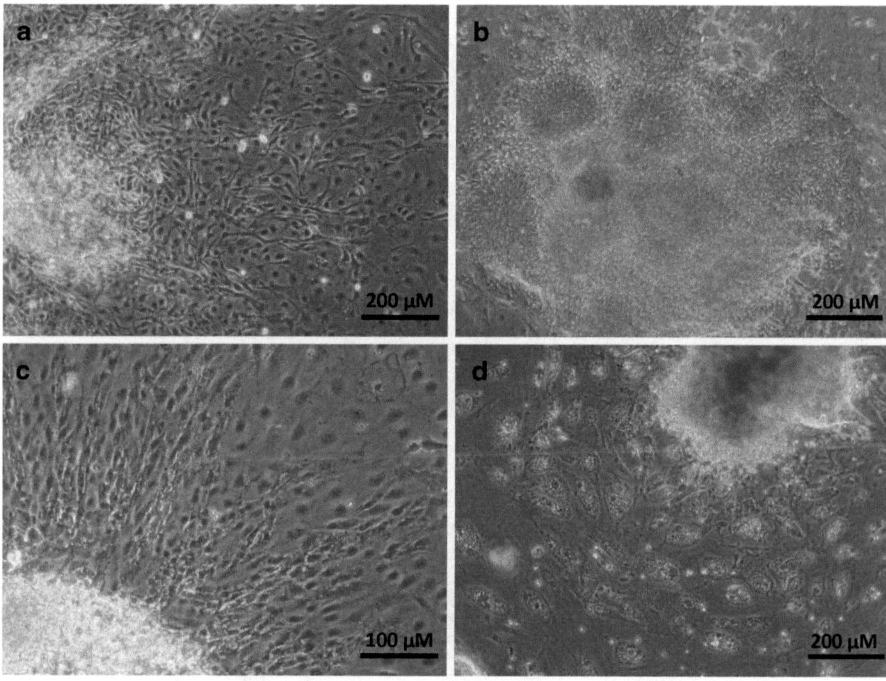

Fig. 5.10 Morphology of cells formed by plated EBs. (**a**) Plated EBs formed by hESCs CL1 with neuron-like morphology. (**b**) Rosette structure formed by plated EBs from hiPSC line hCB110. (**c**) Plated EBs formed by hESCs CL1 with neuron-like morphology. (**d**) Lipid droplet containing cells, formed by plated EBs from hiPSC line hCB1. (**a**, **b**, **d**) Bar 200 μM, (**c**) bar 100 μM

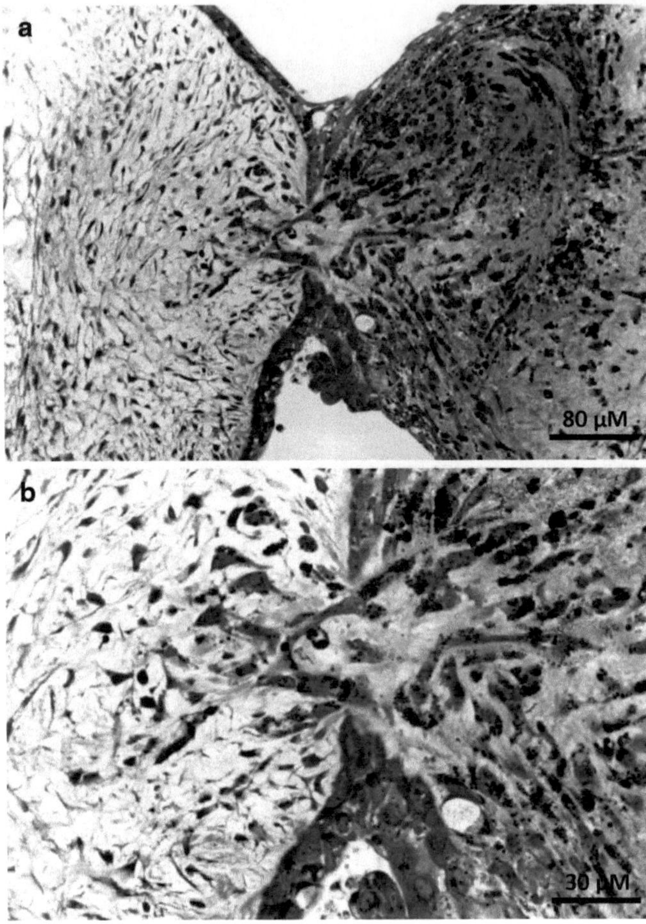

Fig. 5.11 Attached EBs. (**a**) Attached EBs formed by hESC line H9.2, cultured for 14 days. (**b**) Higher magnification of (**a**), in which migration of cells between the attached EBs can be seen. (**a**) Bar 80 μM, (**b**) bar 30 μM. 1-μM Epon histological sections stained with alkaline toluidine blue

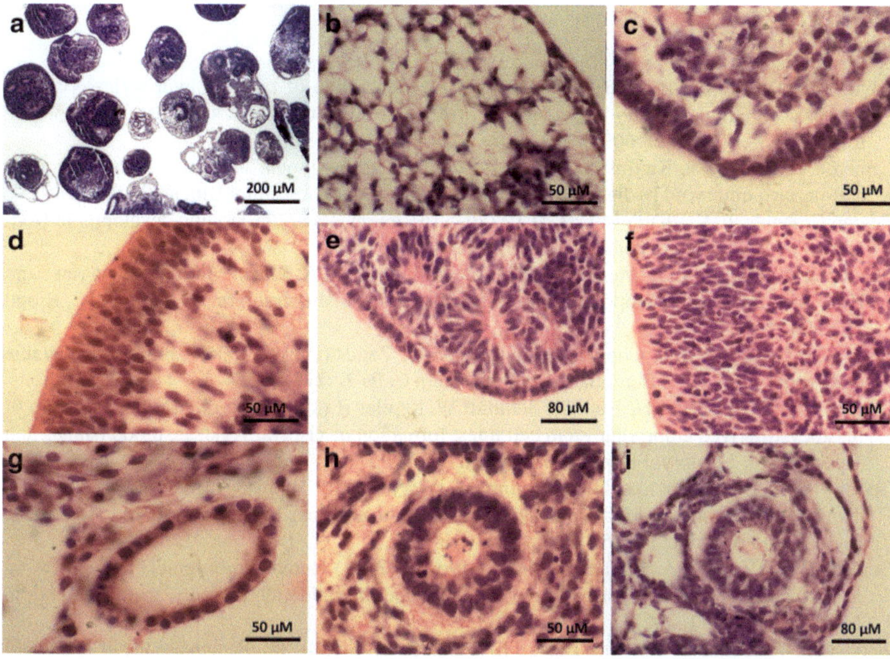

Fig. 5.12 Histological section of EBs cultured in spinner flasks. EBs formed by hESC line H9.2, cultured for up to 14 days in spinner flasks resulting in (**a**) relatively homogenous EBs. (**b**) Undefined differentiating cells resembling adiposities. (**c**) Stratified epithelium at the periphery of EB. (**d**) Rosette structure formed within EBs. (**e, f**) Undefined cells forming unique epithelium lining EBs, resembling neural epithelium. (**g**) Cubal epithelium lining a lumen (**e, f**). Tube-like structures lined by stratified epithelium. (**a**) Bar 200 μM, (**b–d**, **f–h**) bar 50 μM, and (**e**, **i**) bar 80 μM. 1-μM Epon histological sections stained with H&E

References

Buytaert-Hoefen KA, Alvarez E, Freed CR (2004) Generation of tyrosine hydroxylase positive neurons from human embryonic stem cells after coculture with cellular substrates and exposure to GDNF. Stem Cells 22:669–674

Cameron CM, Hu WS, Kaufman DS (2006) Improved development of human embryonic stem cell-derived embryoid bodies by stirred vessel cultivation. Biotechnol Bioeng 94:938–948

Cerdan C, Rouleau A, Bhatia M (2004) VEGF-A165 augments erythropoietic development from human embryonic stem cells. Blood 103:2504–2512

Chadwick K, Wang L, Li L, Menendez P, Murdoch B, Rouleau A, Bhatia M (2003) Cytokines and BMP-4 promote hematopoietic differentiation of human embryonic stem cells. Blood 102:906–915

Dang SM, Gerecht-Nir S, Chen J, Itskovitz-Eldor J, Zandstra PW (2004) Controlled, scalable embryonic stem cell differentiation culture. Stem Cells 22:275–282

Doetschman TC, Eistetter H, Katz M, Schmidt W, Kemler R (1985) The *in vitro* development of blastocyst-derived embryonic stem cell lines: formation of visceral yolk sac, blood islands and myocardium. J Embryol Exp Morphol 87:27–45

Gerami-Naini B, Dovzhenko OV, Durning M, Wegner FH, Thomson JA, Golos TG (2004) Trophoblast differentiation in embryoid bodies derived from human embryonic stem cells. Endocrinology 145:1517–1524

Gerecht-Nir S, Ziskind A, Cohan S, Itskovitz-Eldor J (2003) Human embryonic stem cells as an *in vitro* model for human vascular development and the induction of vascular differentiation. Lab Invest 83:1811–1820

Gerecht-Nir S, Cohan S, Itskovitz-Eldor J (2004) Bioreactor cultivation enhances the efficiency of human embryoid body (hEB) formation and differentiation. Biotechnol Bioeng 86:493–502

Germanguz I, Sedan O, Zeevi-Levin N, Shtreichman R, Barak E, Ziskind A, Eliyahu S, Meiry G, Amit M, Itskovitz-Eldor J, Binah O (2011) Molecular characterization and functional properties of cardiomyocytes derived from human inducible pluripotent stem cells. J Cell Mol Med 15:38–51

Gerrard L, Rodgers L, Cui W (2005) Differentiation of human embryonic stem cells to neural lineages in adherent culture by blocking bone morphogenetic protein signaling. Stem Cells 23:1234–1241

Itskovitz-Eldor J, Schuldiner M, Karsenti D, Eden A, Yanuka O, Amit M, Soreq H, Benvenisty N (2000) Differentiation of human embryonic stem cells into embryoid bodies comprising the three embryonic germ layers. Mol Med 6:88–95

Kaufman DS, Hanson ET, Lewis RL et al (2001) Hematopoietic colony forming cells derived from human embryonic stem cells. Proc Natl Acad Sci USA 98:10716–10721

Kehat I, Kenyagin-Karsenti D, Snir M, Segev H, Amit M, Gepstein A, Livne E, Binah O, Itskovitz-Eldor J, Gepstein L (2001) Human embryonic stem cells can differentiate into myocytes with structural and functional properties of cardiomyocytes. J Clin Invest 108:407–414

Kehat I, Gepstein A, Spira A, Itskovitz-Eldor J, Gepstein L (2002) High-resolution electrophysiological assessment of human embryonic stem cell-derived cardiomyocytes: a novel *in vitro* model for the study of conduction. Circ Res 91:659–661

Koch P, Opitz T, Steinbeck JA, Ladewig J, Brüstle O (2009) A rosette-type, self-renewing human ES cell-derived neural stem cell with potential for *in vitro* instruction and synaptic integration. Proc Natl Acad Sci USA 106:3225–3230

Laflamme MA, Chen KY, NaumovaAV MV, Fugate JA, Dupras SK, Reinecke H, Xu C, Hassanipour M, Police S, O'Sullivan C, Collins L, Chen Y, Minami E, Gill EA, Ueno A, Yuan C, Gold J, Murry CE (2007) Cardiomyocytes derived from human embryonic stem cells in pro-survival factors enhance function of infarcted rat hearts. Nat Biotechnol 25:1015–1024

Levenberg S, Golub JS, Amit M, Itskovits-Eldor J, Langer R (2002) Endothelial cells derived from human embryonic stem cells. Proc Natl Acad Sci USA 99:4391–4396

Mummery C, Ward D, van den Brink CE, Bird SD, Doevendans PA, Opthof T, Brutel de la Riviere A, Tertoolen L, van der Heyden M, Pera M (2002) Cardiomyocyte differentiation of mouse and human embryonic stem cells. J Anat 200:233–242

Mummery C, Ward-van Oostwaard D, Doevendans P, Spijker R, van den Brink S, Hassink R, van der Heyden M, Opthof T, Pera M, de la Riviere AB, Passier R, Tertoolen L (2003) Differentiation of human embryonic stem cells to cardiomyocytes: role of coculture with visceral endoderm-like cells. Circulation 107:2733–2740

Perrier AL, Tabar V, Barberi T, Rubio ME, Bruses J, Topf N, Harrison NL, Studer L (2004) Derivation of midbrain dopamine neurons from human embryonic stem cells. Proc Natl Acad Sci U S A 101:12543–12548

Takahashi K, Yamanaka S (2006) Induction of pluripotent stem cells from mouse embryonic and adult fibroblast cultures by defined factors. Cell 126:663–676

Tanaka T, Tohyama S, Murata M, Nomura F, Kaneko T, Chen H, Hattori F, Egashira T, Seki T, Ohno Y, Koshimizu U, Yuasa S, Ogawa S, Yamanaka S, Yasuda K, Fukuda K (2009) *In vitro* pharmacologic testing using human induced pluripotent stem cell-derived cardiomyocytes. Biochem Biophys Res Commun 385:497–502

Thomson JA, Itskovitz-Eldor J, Shapiro SS, Waknitz MA, Swiergiel JJ, Marshall VS, Jones JM (1998) Embryonic stem cell lines derived from human blastocysts. Science 282:1145–1147 [erratum in Science (1998) 282, 1827]

Vodyanik MA, Bork JA, Thomson JA, Slukvin II (2005) Human embryonic stem cell-derived CD34+ cells: efficient production in the co-culture with OP9 stromal cells and analysis of lymphohematopoietic potential. Blood 105:617–626

Xu C, Police S, Rao N, Carpenter MK (2002a) Characterization and enrichment of cardiomyocytes derived from human embryonic stem cells. Circ Res 91:501–508

Xu RH, Chen X, Li DS, Li R, Addicks GC, Glennon C, Zwaka TP (2002b) Thomson JA BMP4 initiates human embryonic stem cell differentiation to trophoblast. Nat Biotechnol 20: 1261–1264

Yang L, Soonpaa MH, Adler ED, Roepke TK, Kattman SJ, Kennedy M, Henckaerts E, Bonham K, Abbott GW, Linden RM, Field LJ, Keller GM (2008) Human cardiovascular progenitor cells develop from a KDR + embryonic-stem-cell-derived population. Nature 453:524–528

Yirme G, Amit M, Laevsky I, Osenberg S, Itskovitz-Eldor J (2008) Establishing a dynamic process for the formation, propagation, and differentiation of human embryoid bodies. Stem Cells Dev 17:1227–1241

Zhang SC, Wernig M, Duncan ID et al (2001) *In vitro* differentiation of transplantable neural precursors from human embryonic stem cells. Nat Biotechnol 19:1129–1133

Zhang J, Wilson GF, Soerens AG et al (2009) Functional cardiomyocytes derived from human induced pluripotent stem cells. Circ Res 104:e30–e41

Zwi L, Caspi O, Arbel G, Huber I, Gepstein A, Park IH, Gepstein L (2009) Cardiomyocyte differentiation of human induced pluripotent stem cells. Circulation 120:1513–1523

Mummery C, Ward D, van den Brink CE, Bird SD, Doevendans PA, Opthof T, Brutel de la Riviere A, Tertoolen L, van der Heyden M, Pera M (2002) Cardiomyocyte differentiation of mouse and human embryonic stem cells. J Anat 200:233–242

Mummery C, Ward-van Oostwaard D, Doevendans P, Spijker R, van den Brink S, Hassink R, van der Heyden M, Opthof T, Pera M, de la Riviere AB, Passier R, Tertoolen L (2003) Differentiation of human embryonic stem cells to cardiomyocytes [illegible]

[illegible]

Chapter 6
Differentiation of Pluripotent Stem Cells In Vivo: Teratoma Formation

Abstract Human pluripotent cells (hPSCs) are known for their capability of differentiating into cell types representative of the three embryonic germ layers. The pluripotency of both mouse embryonic stem cells (ESCs) and induced pluripotent stem cells (iPSCs) was demonstrated by their capability of contributing to tissue development, including germ cells, after injection into embryos during chimera formation. However, for ethical reasons, this ultimate test of pluripotency cannot be used for hPSCs. Thus, teratoma formation remains one of the most powerful assays for testing hPSC differentiation capabilities. After their injection into several sites in vivo, such as the hindlimb muscle and under the kidney or testis capsule, of severe combined immune deficient (SCID) mice, iPSCs form nonmalignant tumors that contain differentiated cells and tissue formation. In some cases, combinations of cells originating from different germ layers are formed within these teratoma tissues. This chapter focuses on teratoma formation and the tissues that can be indentified within them.

6.1 Introduction

The superiority of the developmental potential of undifferentiated embryonic stem cells (ESCs) over that of adult stem cells or other multipotent stem cells is evident by four main features of mouse ESCs: (a) their consistent contribution to chimera formation, particularly regarding the germ cell line (Bradley et al. 1984), (b) the formation by some lines of entire viable fetuses after their injection into tetraploid embryos (Nagy et al. 1993), (c) their subcutaneous injection into severe combined immune deficient (SCID) mice induces the formation of teratomas containing cells and tissues representative of the three germ layers (Wobus et al. 1984), and (d) their transfer to suspension culture in vitro induces the spontaneous

M. Amit and J. Itskovitz-Eldor, *Atlas of Human Pluripotent Stem Cells: Derivation and Culturing*, Stem Cell Biology and Regenerative Medicine,
DOI 10.1007/978-1-61779-548-0_6, © Springer Science+Business Media, LLC 2012

formation of cell aggregations known as embryoid bodies (EBs), with areas differentiating into embryonically distinct cell types (Doetschman et al. 1985). Mouse induced pluripotent stem cells (iPSCs) demonstrated these same features, including chimera contribution and formation of entire viable fetuses (Takahashi and Yamanaka 2006; Meissner et al. 2007). However, for human pluripotent cells (hPSCs), chimera or entire viable fetus formation is impossible, due to technical and ethical difficulties, leaving teratoma formation as the most powerful tool for demonstrating cell pluripotency.

Teratomas are unique benign tumors containing cell types and tissues representative of the three embryonic germ cells. Teratomas are affiliated with the class of germ cell tumors (Blum and Benvenisty 2008). The name "teratoma" comes from the Latin word *Teratos* (a "monster"), since the tissues within the tumor are laid down with no order, and in some incidences, a tooth, hair or other small organ appears. According to Blum and his colleagues, pathological examination does not enable detection of undifferentiated clusters within teratomas, or of cells that still express *Oct4* or Nanog (Blum et al. 2009; Blum and Benvenisty 2007). Furthermore, while teratomas are regarded as polyclonal tumors developed from karyotypically normal cells, teratocarcinoma development starts from clonal origin and the tumor consists of abnormal cells (Blum et al. 2009). Nevertheless, PSCs undergoing genetic or epigenetic changes in vitro may form malignant teratocarcinoma (Blum and Benvenisty 2009).

Teratoma can be induced by injecting PSCs into several sites in mice. It seems that the site of injection can affect teratoma formation rates and growth (Prokhorova et al. 2009). Prokhorova and his colleagues found kidney capsule injection to result in the highest success rates (100%), followed by subcutaneous injection using Matrigel (80–100%), intramuscular injection (60%) and intramuscular injection (12.5%). However, using SCID beige mice and high cell numbers (of at least 6×10^6 cells per mice), we have achieved success rates of up to 80% for teratoma formation by intramuscular injection, thus circumventing the need for operating on the mice and for using sedatives (Amit et al. unpublished data).

The formation of teratomas is regarded as a disadvantage of using PSCs for cell-based therapy (Blum and Benvenisty 2008). Lee and his colleagues demonstrated that contamination of 10^4 undifferentiated cells injected, together with mature differentiated cells, into skeletal muscle was sufficient to cause teratoma formation (Lee et al. 2009). Therefore, to exclude the possibility of undifferentiated cell contamination, followed by the risk of teratoma formation, a long safety test of at least 12 weeks should be conducted (Fong et al. 2010).

6.2 Materials for Teratoma Formation

6.2.1 Culture Medium

There is no need for a special medium; use the conditioned medium for culturing the injected cells.

6.2.2 Syringe for Injecting Cells

Use 1-ml syringe (insulin syringe) with an 18–25-gauge needle.

6.3 Formation of Teratomas

6.3.1 Protocol for Teratoma Formation

1. Scrap hESCs drawn from three confluent wells (out of a 6-well plate, 30 cm^2, 6×10^6–10^7 cells) using a policeman rubber scraper. Collect cells into a 4-ml tube or an Eppendorff tube (to allow collecting all cell pellets into the syringe).
2. Centrifuge the cells for 5 min at $90 \times g$.
3. Leave as little medium as possible, and upload the cell pellets into the syringe. There is no need to break the cell pellets.
4. Inject the cells into the rear leg muscles of 4-week-old male SCID beige mice, using a 18- or 21-gauge needle.
5. Examine the resultant teratoma 10 weeks after injection.

6.3.2 Routine Treatment of Mice and Teratoma

Higher concentrations of injected cells will ensure teratoma formation but will reduce the mice's capability to carry the tumors. About 6 weeks after injection, teratoma formation can be detected by touch. At 10–12 weeks after injection, the teratoma can be harvested from mice. The tumor tissue can be used for RNA, DNA, or protein isolation using standard protocols. The teratoma tissue should be fixed immediately after harvest, using a standard fixer for paraffin-embedded sections (based on 10% paraformaldehyde) or by preparation of frozen sections.

In mature teratoma, a variety of tissue types can be seen (Fig. 6.1). In some sections, tissue representative of the three embryonic germ layers can be identified in the same field (Fig. 6.1c). Ectoderm-originated tissues formed in teratoma may

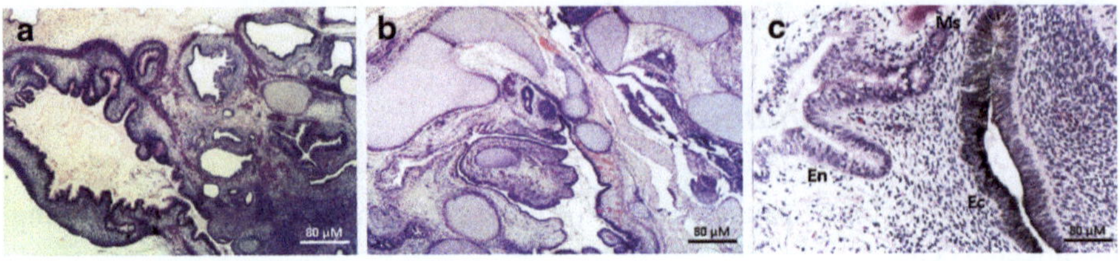

Fig. 6.1 Low-magnification section of teratoma. View of teratoma histological sections showing a variety of tissues formed by (**a**) pluripotent cell line J3 and (**b**) hESC line I6. (**c**) A section containing representative tissues of the three embryonic germ layer formed by hESC line WS1: ectoderm, epithelium containing melanin-producing cells (marked Ec); mesoderm, bone tissue (marked Ms); and endoderm, columnar epithelium containing goblet cells (marked En). (**a**) and (**b**) Bar 750 μM, and (**c**) Bar 150 μM. 1-μM Epon histological sections stained with hematoxylin & eosin (H&E)

include stratified epithelium (Fig. 6.2), neural epithelium with melanin containing cells (Fig. 6.3), and nerve fibers (Fig. 6.4). Examples of mesoderm-originated tissues include fat (Fig. 6.5), muscle (Fig. 6.6), cartilage (Fig. 6.7), and bone (Fig. 6.8). From endoderm-originated tissues, columnar epithelium containing goblet cells, gland structure, and embryonic-glomerulus-like structures can be seen (Fig. 6.9). In some teratoma, complex organs, such as hair and hair follicles (Fig. 6.10), eye formation, bone tissue containing bone marrow, and skin formation (Fig. 6.11) are apparent.

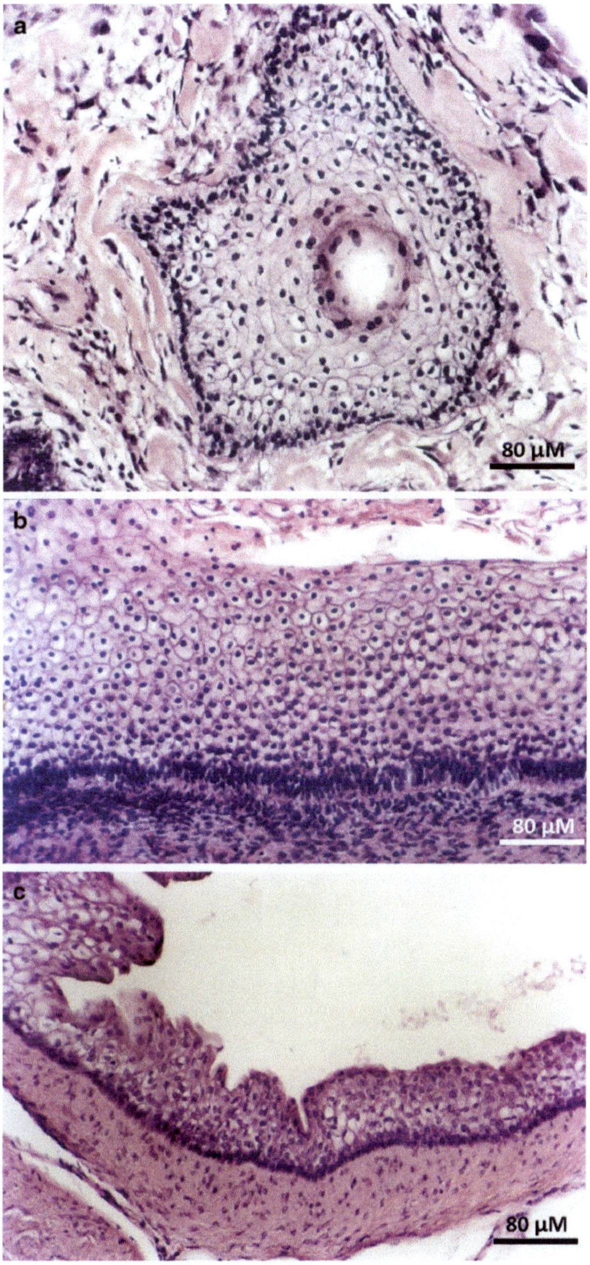

Fig. 6.2 Tissue originated from ectoderm—stratified epithelium of ectoderm origin. Examples of stratified epithelium of ectodermal origin formed by (**a**) hESC line WS1, (**b**) hESC line I3, and (**c**) hESC line I6, which was cultured under feeder layer-free conditions. (**a**) Bar 90 μM, (**b**) and (**c**) Bar 150 μM. 1-μM Epon histological sections stained with H&E

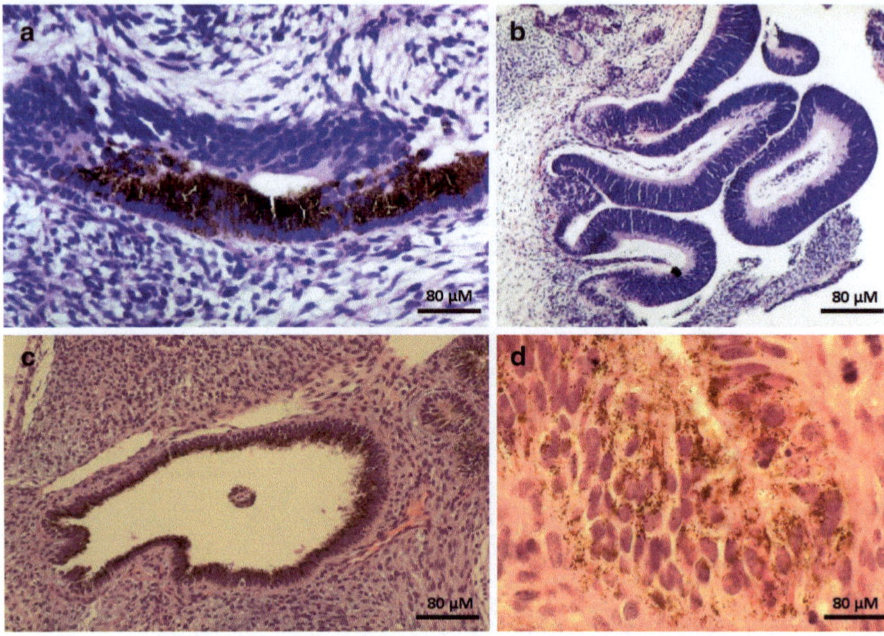

Fig. 6.3 Tissue originated from ectoderm—epithelium of ectoderm origin containing melanin-producing cells. (**a**) Examples of epithelium of ectodermal origin consisting of melanin-producing cells formed by the pluripotent cell line J3. (**b**, **c**) An example of epithelium without melanin-producing cells formed by (**b**) hESC line I6 (**c**) hESC line I3 which was cultured in suspension for 26 passages. (**d**) The same epithelium of (**c**), at higher magnification. (**a**) Bar 90 μM, (**b**) Bar 350 μM, (**c**) Bar 150 μM, and (**d**) Bar 30 μM. 1-μM Epon histological sections stained with H&E

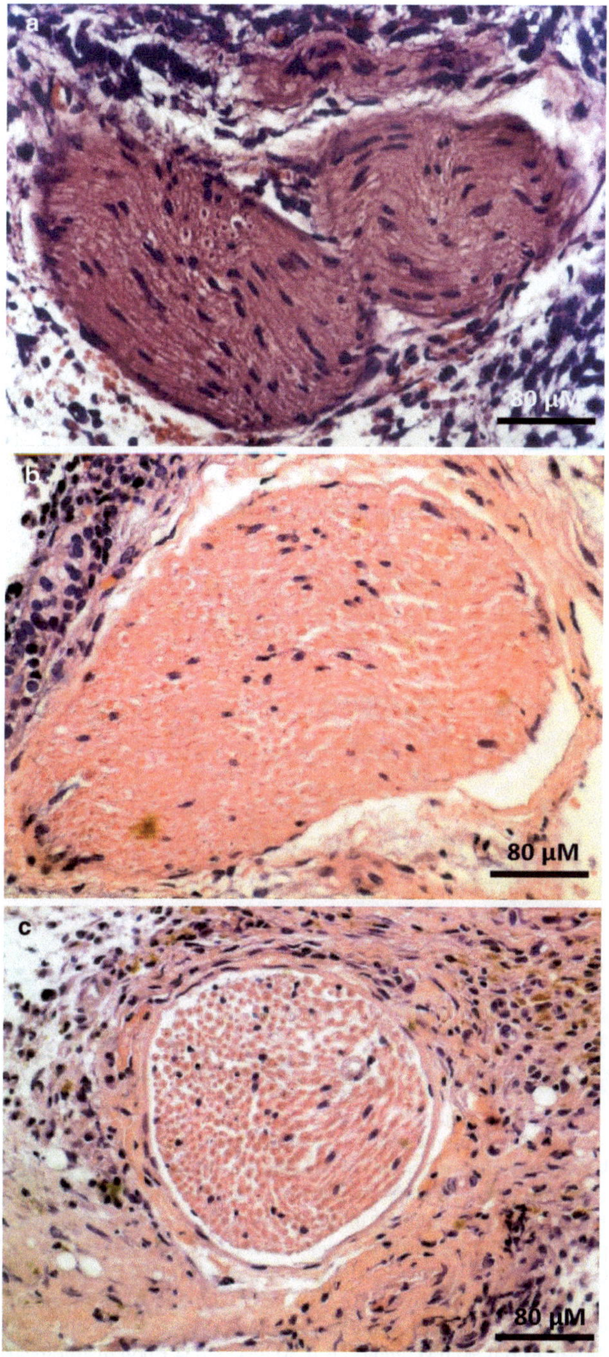

Fig. 6.4 Tissue originated from ectoderm—myelinated nerve. Examples of myelinated nerve fibers formed by (**a**) hESC line I3, (**b**) and (**c**) hESC line I4, which were cultured in suspension for 9 passages. In each fiber, a darker axon is apparent, surrounded by a *pink* myelin envelop. Bar 50 μM. 1-μM Epon histological sections stained with H&E

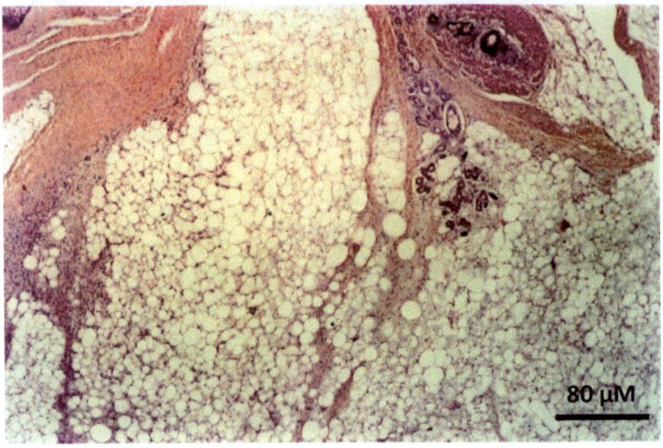

Fig. 6.5 Tissues originated from mesoderm—fat tissue. Example of fat tissue formed by pluripotent cell line J3. Bar 300 µM. 1-µM Epon histological sections stained with H&E

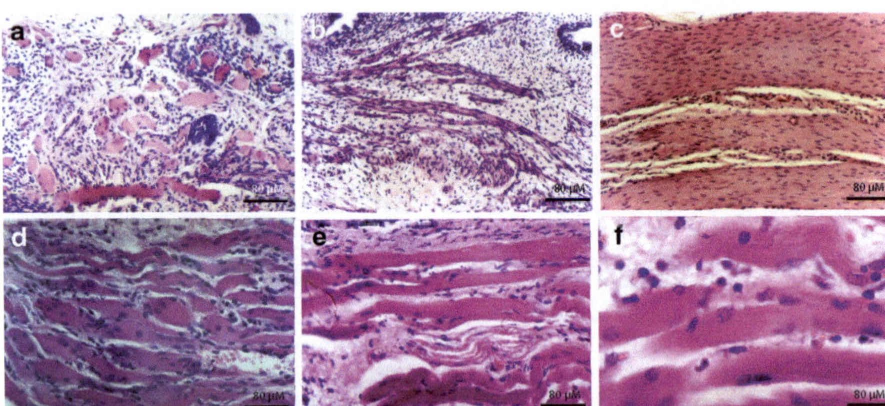

Fig. 6.6 Tissues originated from mesoderm—muscle tissue. Example of muscle tissue formed in teratoma. (**a**) Skeletal muscle formed by hESC line I6, transverse section. (**b**) Smooth muscle formed by hESC line I6 cultured under feeder layer-free conditions. (**c**) Wide tissue of smooth muscle formed by hESC line I6 cultured under feeder layer-free conditions. (**d**) Skeletal muscle formed by pluripotent cell line J3, partly transverse and partly vertical section. (**e**) Skeletal muscle formed by hESC cell line H9, cultured with human foreskin fibroblasts as feeder layer, vertical section. (**f**) Skeletal muscle formed by hESC cell line I6, cultured with human foreskin fibroblasts as feeder layer, vertical section. In this section, the typical strip formed by the Z-bands is apparent. (**a–c**) Bar 150 µM, (**d, e**) Bar 90 µM, and (**f**) Bar 30 µM. 1-µM Epon histological sections stained with H&E

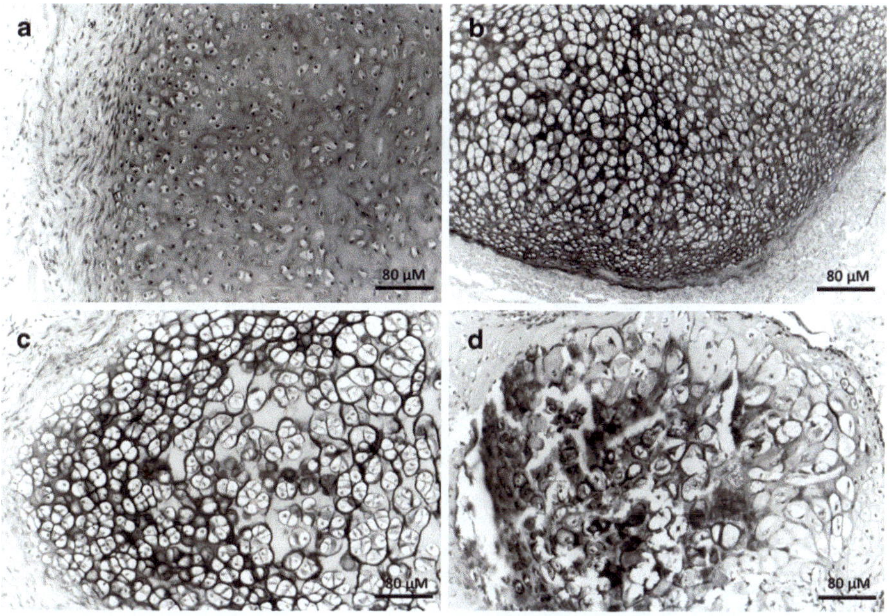

Fig. 6.7 Tissues originated from mesoderm—cartilage tissue. Example of cartilage development in teratoma formed by hESC line H9.2. (**a**) Immature tissue. (**b**) Mature hyaline cartilage. (**c**) An additional example of mature hyaline cartilage. (**d**) Calcified cartilage. Bar 50 μM. 1-μM Epon histological sections stained with H&E

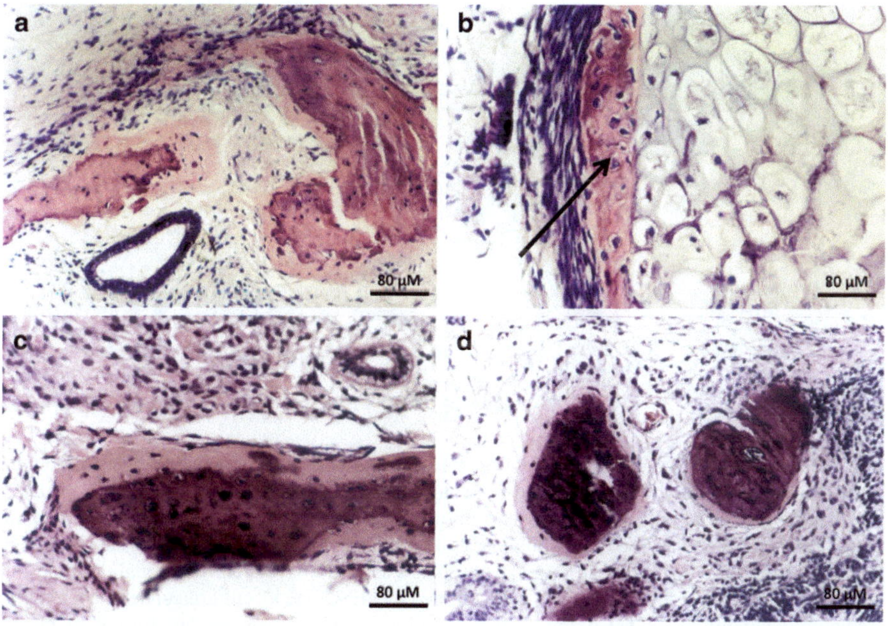

Fig. 6.8 Tissues originated from mesoderm—bone tissue. (**a**) An example of bone tissue formed in teratoma by pluripotent cell line J3. (**b**) Bone ossification on cartilage tissue formed by pluripotent cell line J3 (the bone is marked with a *black arrow*). (**c**) and (**d**) Additional examples of bone formation by hESC line WS1. (**a**, **c**, **d**) Bar 150 μM, (**b**) Bar 30 μM. 1-μM Epon histological sections stained with H&E

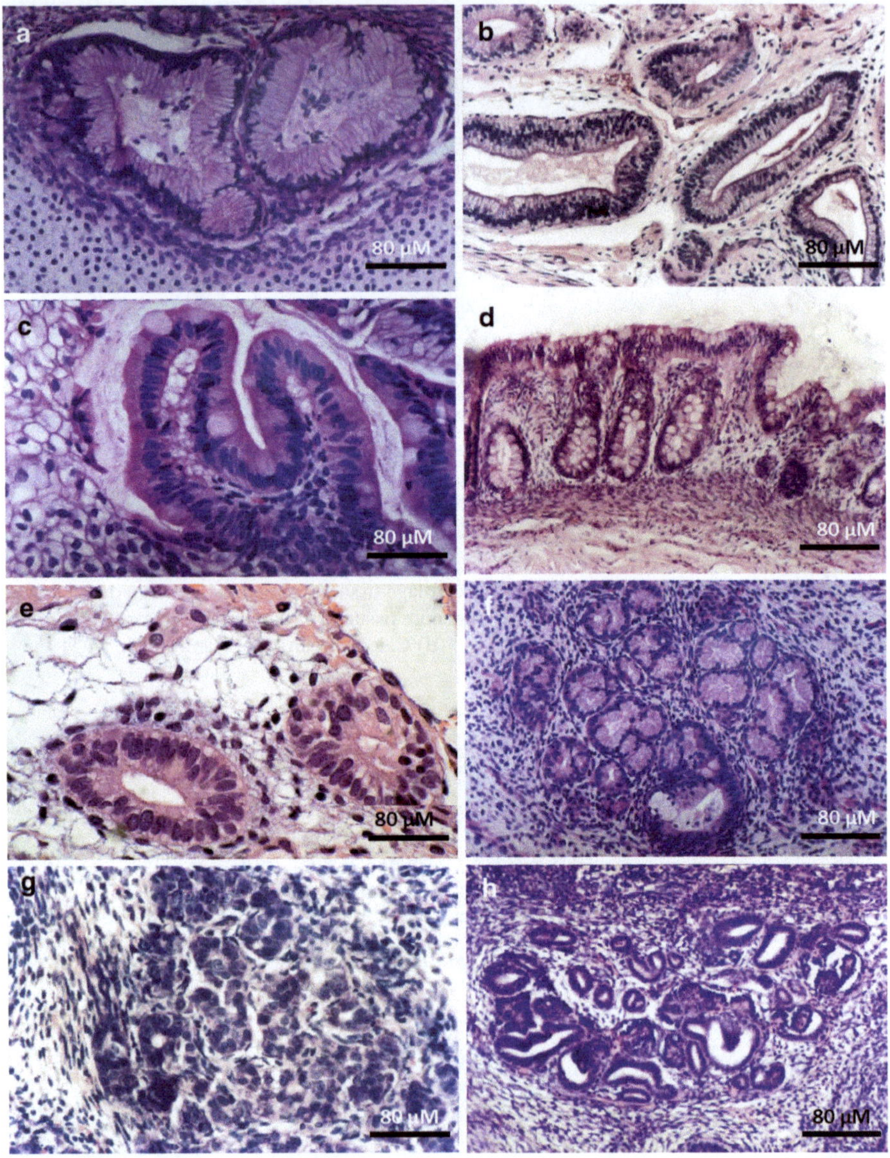

Fig. 6.9 Tissues originated from endoderm—tissues from endoderm origin. Examples of epithelial tissue from endoderm origin formed in teratoma. (**a**) Columnar epithelium formed by hESC cell line H9, cultured with human foreskin fibroblasts as feeder layer. (**b**) Stratified epithelium formed by hESC cell line WS1. (**c**) Columnar epithelium containing goblet cells, formed by hESC cell line H9, cultured with human foreskin fibroblasts as feeder layer. (**d**) Additional examples of columnar epithelium containing goblet cells, formed by hESC line I6, cultured under feeder layer-free culture conditions. (**e, f**) Columnar epithelium lining a gland-like structure formed by hESC line H9, cultured with human foreskin fibroblasts as feeder layer. (**g**) Cubical epithelium lining a gland-like structure formed by hESC line I6, cultured under feeder layer-free culture conditions. (**h**) Embryonic glomeruli-like structures formed by hESC line I6. (**a, c, e, g**) Bar 90 μM, (**b, d, f, h**) Bar 150 μM. 1-μM Epon histological sections stained with H&E

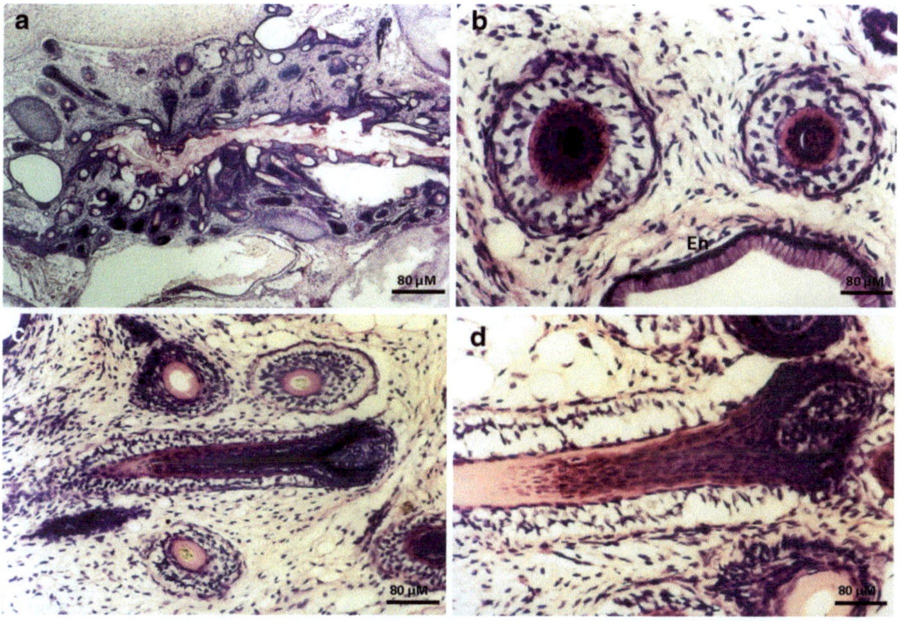

Fig. 6.10 Complex tissue and organ formation—hair formation. Hair follicles and hair formed by hESC line H9.2 in teratoma. (**a**) Area in teratoma containing hair and hair follicles. (**b**) Two hair follicles (a columnar epithelium can also be seen). (**c**) A strand of hair and a few hair follicles (**d**) The strand of hair at high magnification. (**a**) Bar 750 μM, (**b**, **d**) Bar 50 μM, and (**c**) Bar 150 μM. 1-μM Epon histological sections stained with H&E

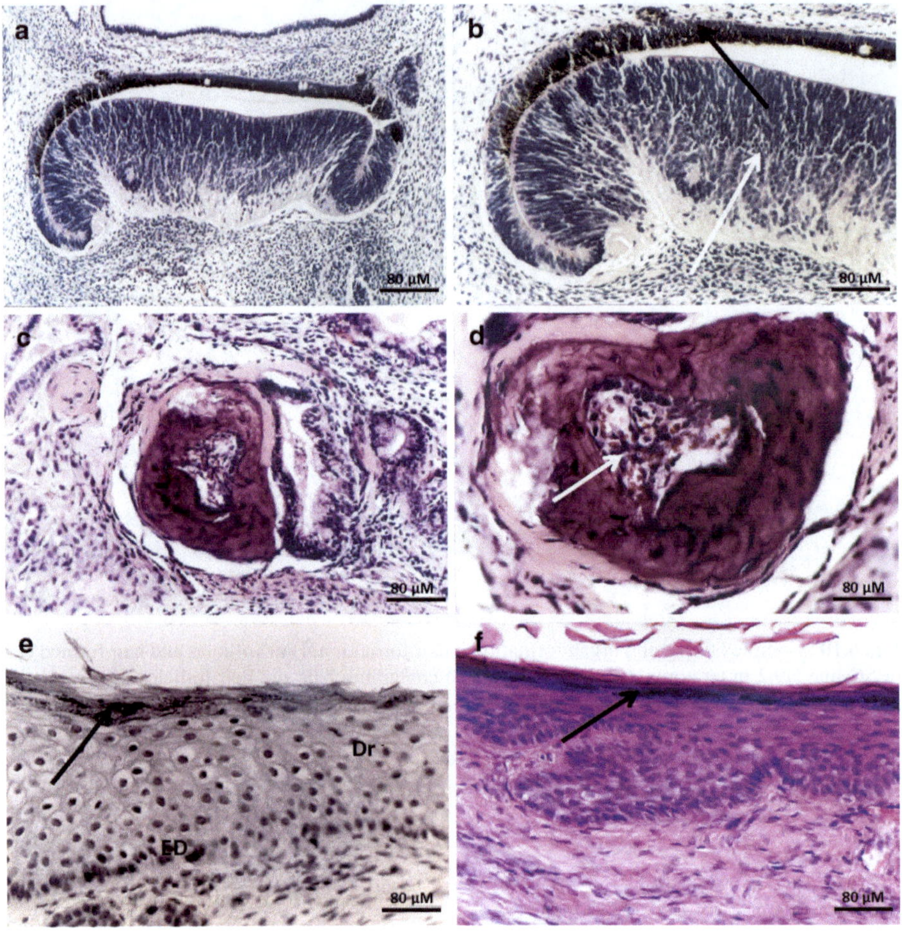

Fig. 6.11 Complex tissue and organ formation—organs formed in teratoma. Examples of complex organs formed in teratoma. (**a**) Eye formation, including retina and lance formation, formed by the hESC line WS1. (**b**) High magnification of (**a**): The retina is marked by a *black arrow* and the lance by a *white arrow*. (**c**) Bone tissue containing bone marrow cells formed by hESC line WS1. (**d**) High magnification of (**c**), the bone marrow cells are marked by a *white arrow*. (**e, f**) Skin formation by hESC lines H9.2 and H9 cultured with human foreskin fibroblasts, for (**e**) and (**f**), respectively. The stratum corneum containing keratin cells is marked by a *black arrow*, the dermis is marked **Dr**, and the stratum basale by **SB**. (**a, c**) Bar 90 μM, (**b–f**) Bar 30 μM. 1-μM Epon histological sections stained with H&E, (**e**) was stained with alkaline toluidine blue

References

Blum B, Benvenisty N (2007) Clonal analysis of human embryonic stem cell differentiation into teratomas. Stem Cells 25:1924–1930

Blum B, Benvenisty N (2008) The tumorigenicity of human embryonic stem cells. Adv Cancer Res 100:133–158

Blum B, Benvenisty N (2009) The tumorigenicity of diploid and aneuploid human pluripotent stem cells. Cell Cycle 8:3822–3830

Blum B, Bar-Nur O, Golan-Lev T, Benvenisty N (2009) The anti-apoptotic gene survivin contributes to teratoma formation by human embryonic stem cells. Nat Biotechnol 27:281–287

Bradley A, Evans M, Kaufman MH, Robertson E (1984) Formation of germ-line chimaeras from embryo-derived teratocarcinoma cell lines. Nature 309:255–256

Doetschman TC, Eistetter H, Katz M, Schmidt W, Kemler R (1985) The *in vitro* development of blastocyst-derived embryonic stem cell lines: formation of visceral yolk sac, blood islands and myocardium. J Embryol Exp Morphol 87:27–45

Fong CY, Gauthaman K, Bongso A (2010) Teratomas from pluripotent stem cells: a clinical hurdle. Cell Biochem 111:769–781

Lee AS, Tang C, Cao F, Xie X, van der Bogt K, Hwang A, Connolly AJ, Robbins RC, Wu JC (2009) Effects of cell number on teratoma formation by human embryonic stem cells. Cell Cycle 8:2608–2612

Meissner A, Wernig M, Jaenisch R (2007) Direct reprogramming of genetically unmodified fibroblasts into pluripotent stem cells. Nat Biotechnol 25:1177–1181

Nagy A, Rossant J, Nagy R, Abramow-Newerly W, Roder JC (1993) Derivation of completely cell culture-derived mice from early-passage embryonic stem cells. Proc Natl Acad Sci USA 90: 8424–8428

Prokhorova TA, Harkness LM, Frandsen U, Ditzel N, Schrøder HD, Burns JS, Kassem M (2009) Teratoma formation by human embryonic stem cells is site dependent and enhanced by the presence of Matrigel. Stem Cells Dev 18:47–54

Takahashi K, Yamanaka S (2006) Induction of pluripotent stem cells from mouse embryonic and adult fibroblast cultures by defined factors. Cell 126:663–676

Wobus AM, Holzhausen H, Jakel P, Schoneich J (1984) Characterization of pluripotent stem cell line derived from mouse embryo. Exp Cell Res 152:212–219

Chapter 7
Immunostaining

Ilana Laevsky

Abstract Immunostaining refers to the processes of localizing antigens in cells, thus exploiting the principle that antibodies bind specifically to antigens in biological tissues. There are two general methods of immunostaining: immunohistochemistry (IHC) and immunofluorescense. For human embryonic stem cells (hESCs), researchers use immunostaining to detect specific undifferentiation markers, to demonstrate pluripotency by embryoid body (EB) and by teratoma formation (immunostaining with three germ layer markers), and to detect specific differentiation processes.

7.1 Introduction

Immunohistochemistry (IHC) is the method of in situ detection of antigens in cells and tissue sections (formalin fixed, paraffin embedded, frozen, etc.) by specific antibodies. Most commonly, an antibody is conjugated to an enzyme, such as peroxidase, that can catalyze a color-producing reaction. Alternatively, an antibody can be tagged to a fluorophore, such as fluorescein or FITC (immunofluorescence, IF).

Antibodies used for specific detections can be monoclonal or polyclonal. Monoclonal antibodies generally exhibit greater specificity. Polyclonal antibodies are a heterogeneous mix of antibodies that recognize several epitopes. Antibodies

I. Laevsky, B.A. (✉)
Stem Cell Research Center, The Rappaport Faculty of Medicine,
Technion—Israel Institute of Technology, Haifa, Israel

Department of Obstetrics and Gynecology, Rambam Health Care Campus, Haifa, Israel
e-mail: ilanala@tx.technion.ac.il

M. Amit and J. Itskovitz-Eldor, *Atlas of Human Pluripotent Stem Cells: Derivation and Culturing*, Stem Cell Biology and Regenerative Medicine,
DOI 10.1007/978-1-61779-548-0_7, © Springer Science+Business Media, LLC 2012

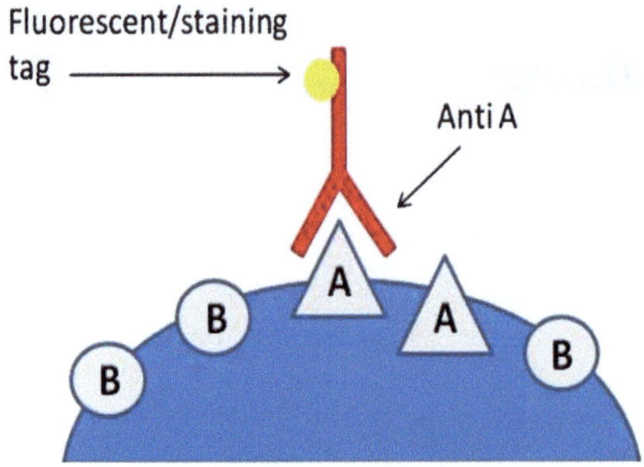

Fig. 7.1 The direct method of immunohistochemical labeling uses one labeled antibody, which binds directly to the antigen being targeted

can also be classified as primary or secondary reagents. Primary antibodies are raised against an antigen of interest and are typically unconjugated (unlabeled), while secondary antibodies are raised against primary antibodies. Hence, secondary antibodies recognize immunoglobulins of a particular species and are conjugated to either biotin or a reporter enzyme, such as alkaline phosphatase or horseradish peroxidase (HRP). Some secondary antibodies are conjugated to fluorescent agents, such as Alexa Fluor or Cy3 (immunofluorescence). Antigen–antibody hybridization enables detection of an antigen by means of light or fluorescent microscopy (immunofluorescence). Protein concentration is generally measured by densitometry analysis, with the intensity of staining correlating to the amount of the protein investigated.

Two strategies are used for the immunohistochemical detection of antigens in tissue: the primary (direct) method and the secondary (indirect) method. The former consists of a one-step staining protocol involving the direct reaction of a labeled antibody with the antigen in the tissue (Fig. 7.1). This technique is simple and rapid due to the use of only one antibody. However, since sensitivity can be reduced due to low signal amplification, this technique is less commonly used than indirect methods.

The indirect method uses two antibodies. The first (primary) antibody recognizes the target molecule and binds to it. The second (secondary) antibody, which carries the fluorophore or the enzyme, recognizes the primary antibody and binds to it (Fig. 7.2). This protocol is more complex and time consuming than the direct protocol, but is more flexible. The great possibilities of primary and secondary antibodies enable simultaneous detection of various antigens. Nevertheless, immunostaining techniques are generally limited to fixed (i.e., dead) samples, since antibodies in live cells cannot cross the cell membrane.

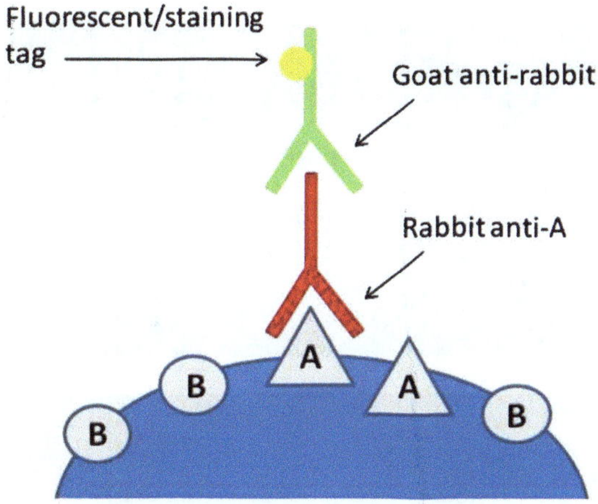

Fig. 7.2 The indirect method of immunohistochemical labeling uses one antibody against the antigen being probed for, and a second labeled antibody against the first

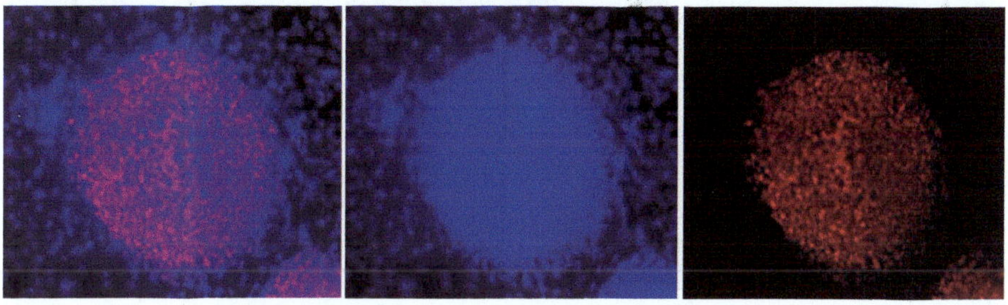

Fig. 7.3 IF: Undifferentiated I3.2 hESCs on MEF marked with OCT4 as a primary antibody. Cy3 is the *red*-colored secondary antibody, and DAPI is the *blue*-colored counterstain

Immunostaining methods can be used in hESC research for a number of purposes: directed differentiation experiments for the analysis of differentiation products; analysis of undifferentiated hESCs (markers such as OCT4, SSEA4, TRA-60, or TRA-81) (Figs. 7.3–7.4); and the demonstration of cell pluripotency in vitro or in vivo by immunostaining of EBs or teratomas (formalin fixed and paraffin embedded, or paraformaldehyde fixated, depending on the case) with antibodies that represent the three germ layers (Figs. 7.5–7.9).

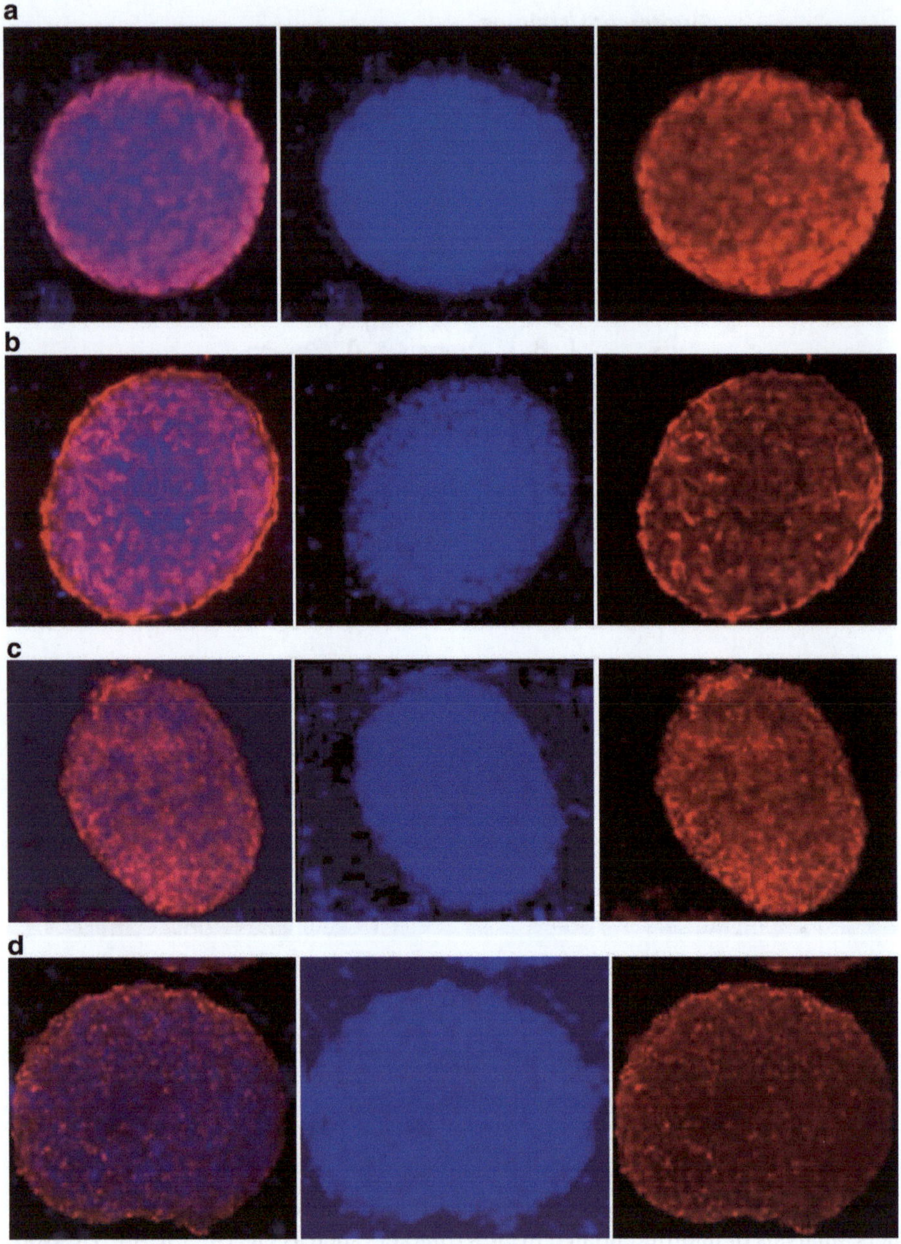

Fig. 7.4 IF: Undifferentiated J1.2.3 iPSCs in suspension marked with OCT4 (**a**), TRA-60 (**b**), SSEA4 (**c**), and SSEA3 (**d**) as primary antibodies. Cy3 is the *red*-colored secondary antibody, and DAPI is the *blue*-colored counterstain

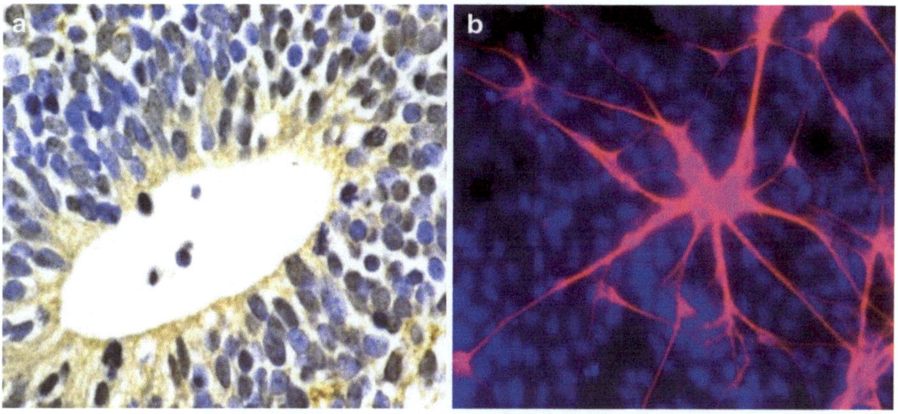

Fig. 7.5 (**a**) IHC: Paraffin-embedded EB preparates from H9.2 hESCs: b-tubulin III is the primary antibody, and the neural rosettes are positively marked; (**b**) IF: I6 hESCs differentiated into the neural lineage: b-tubulin III is the primary antibody, Cy3 is the *red*-colored secondary antibody, and DAPI is the *blue*-colored counterstain

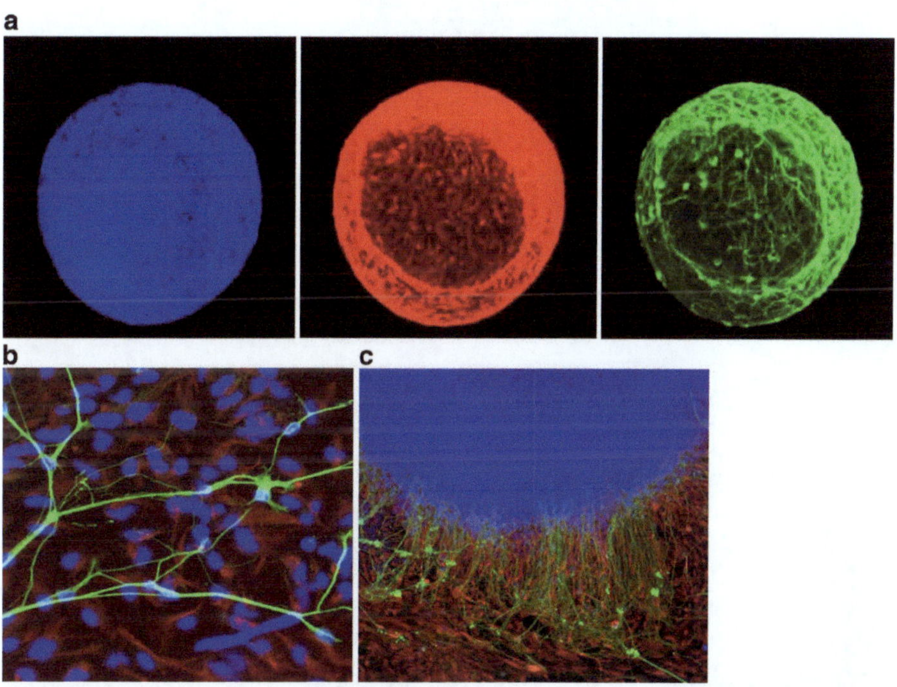

Fig. 7.6 (**a**) IF: iPSC line hCB1 differentiated into the neural lineage: b-tubulin III (*green*) and nestin (*red*) are the primary antibodies, Cy3 is the *red*-colored secondary antibody, alexa488 is the *green*-colored secondary antibody, and DAPI is the *blue* colored counter stain, (**a**) growing in suspension, (**b**) plated on HPF, (**c**) plated EBs

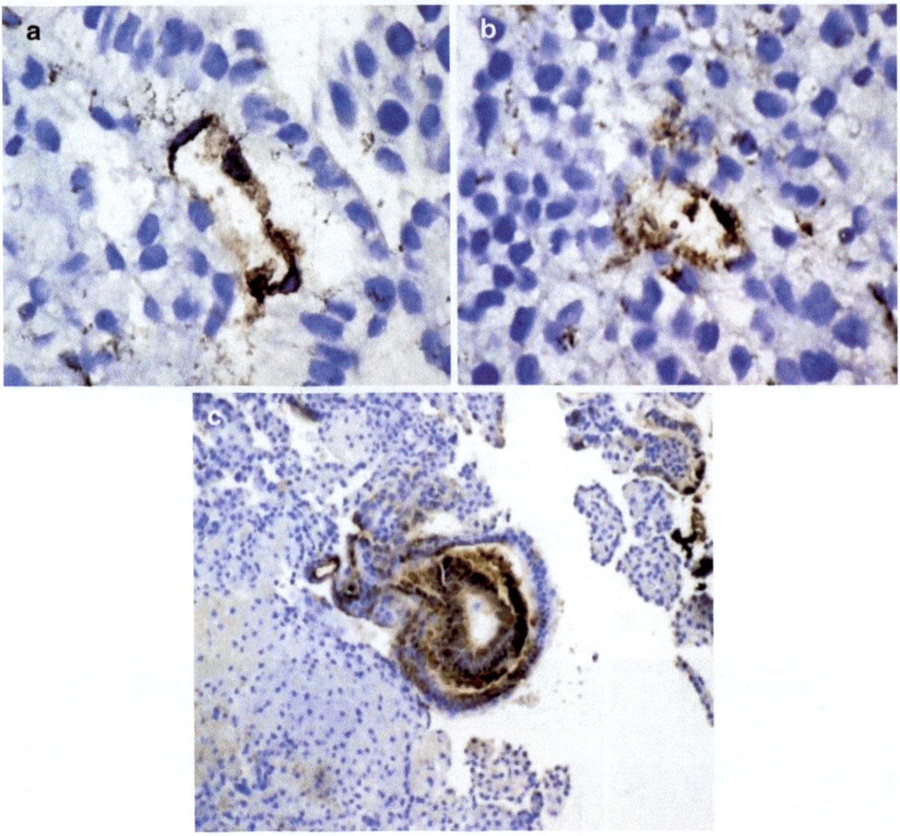

Fig. 7.7 (**a**) IHC: Paraffin-embedded EB preparates: CD-34 (**a, b**), AFP (**c**). As a primary antibody, and the relevant structures are positively marked

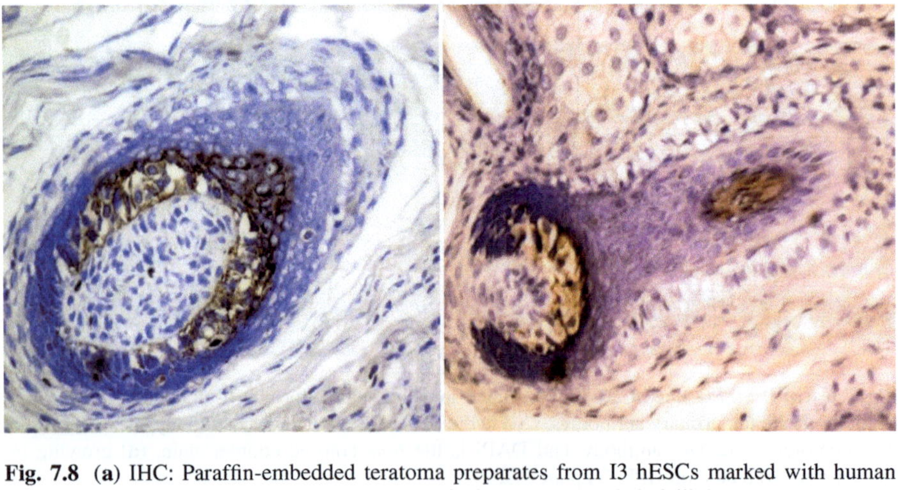

Fig. 7.8 (**a**) IHC: Paraffin-embedded teratoma preparates from I3 hESCs marked with human lamin A/C as a primary antibody. The photographed structures are hair follicles

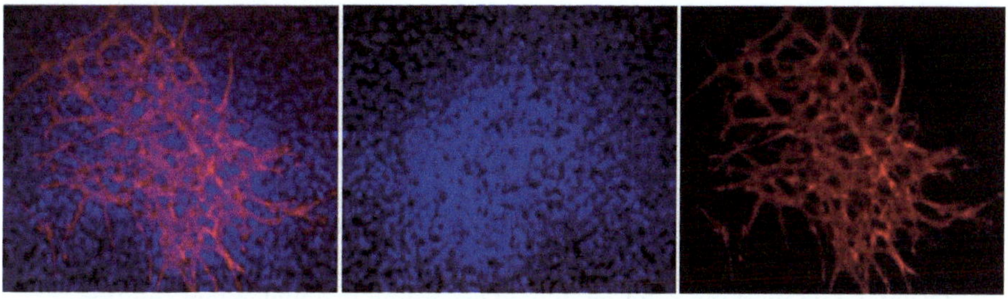

Fig. 7.9 IF: I3 hESCs differentiated into endothelial lineage and marked with CD-31 as a primary antibody. Cy3 is the *red*-colored secondary antibody, and DAPI is the *blue*-colored counterstain

7.2 Materials and Solutions for Immunostaining

7.2.1 Materials and Solutions for Immunohistochemistry of Paraffin-Embedded Tissues

1. Xylene for deparaffinization.
2. Ethanol absolute (100%), ethanol 95%—as a step in the rehydration and dehydration process.
3. H_2O_2 3% solution in methanol for endogenous peroxidase blocking.
4. Phosphate-buffered saline (PBS) for washing.
5. Primary antibody of the researcher's choice.
6. DAKO LSAB+System, HRP, cat. K0690 (includes LINK and streptavidin HRP).
7. DAKO liquid DAB+substrate chromogen system, cat. K3468 (includes buffered substrate for liquid DAB and liquid DAB chromogen).
8. DAKO hematoxylin (HEMA), cat. 2020.
9. Zymed antibody diluent, cat. 003118.
10. Mounting medium (Merck's Entallan or other).

7.2.2 Materials and Solutions for Immunofluorescence

1. Paraformaldehyde 16% solution, Electron Microscopy Sciences, cat. 15710.
2. Paraformaldehyde 4% in PBS solution—fixation solution.
3. TritonX-100, Sigma-Aldrich, cat. 114K0182.
4. Triton X-100 1% in PBS—permeabilization solution (which reduces surface tension, thereby improving penetration of the antibodies into the cell).
5. Normal goat serum (NGS) 10%, Invitrogen Corporation, cat. 14190.
6. Triton X-100 1%+NGS 2% in PBS-blocking solution (which reduces nonspecific background staining).

7. Primary antibody of the researcher's choice.
8. Primary antibody working solution in blocking solution, according to a previously validated concentration.
9. Secondary antibody of the researcher's choice (Cy3, Alexa Fluor, etc.).
10. Secondary antibody working solution—1:50 in PBS or according to primary antibody (prepare at room temperature in darkness).
11. Counterstain solution (DAPI, propidium iodine, To-Pro, etc.) of the researcher's choice.
12. Counterstain working solution diluted according to counterstain solution.

7.3 Immunostaining Procedures

7.3.1 Immunohistochemistry of Paraffin-Embedded Tissues

1. Wash slides in xylene for 3 min (×2).
2. Wash slides in ethanol 100% for 1 min (×2, prepare a new solution for each step).
3. Wash slides in ethanol 95% for 1 min (×2, prepare a new solution for each step).
4. Wash slides in tap water for 10 min.
5. Rinse slides in PBS solution.
6. Incubate slides with 3% H_2O_2 in methanol solution for 5 min.
7. Rinse slides in PBS.
8. Incubate slides with primary antibody working solution for 1 h (prepare at room temperature in darkness).
9. Rinse slides in PBS.
10. Incubate slides with LINK for 30 min (prepare at room temperature in darkness).
11. Rinse slides in PBS.
12. Incubate slides with streptavidin peroxidase for 30 min (prepare at room temperature in darkness).
13. Rinse slides in PBS.
14. Prepare substrate chromogen solution: 1 ml buffer + 1 drop DAB chromogen.
15. Incubate slides with the substrate chromogen solution for 3–5 min (prepare at room temperature in darkness).
16. Rinse slides in distilled water.
17. Incubate slides with hematoxylin for 1 min (prepare at room temperature in darkness).
18. Wash slides in tap water for 30 s.
19. Wash slides in ethanol 95% for 1 min (×2, prepare a new solution for each step).
20. Wash slides in ethanol 100% for 1 min (×2, prepare a new solution for each step).

21. Wash slides in xylene for 2 min (×2).
22. Place 1 drop of mounting medium on a preparate and close with a cover slide.
23. Wait overnight for the slides to dry.
24. Analyze staining results with an upright microscope.

7.3.2 *Immunofluorescence of Cultured Cells*

1. Remove medium from the cells.
2. Add fixation solution (see Sect. 7.2.2, step 2) and incubate at room temperature for 30 min.
3. Remove the fixation solution.
4. Rinse with PBS (×2).
5. Add permeabilization solution (see Sect. 7.2.2, step 4) and incubate at room temperature for 10–15 min.
6. Remove the permeabilization solution.
7. Add blocking solution (see Sect. 7.2.2, step 6) and incubate at room temperature for 15–20 min.
8. Remove the blocking solution (excluding the negative control well).
9. Add primary antibody working solution (see Sect. 7.2.2, step 8) (blocking solution for negative control) and incubate overnight at 4°C.
10. Remove the primary antibody working solution.
11. Rinse with PBS (×3).
12. Add secondary antibody working solution (see Sect. 7.2.2, step 10) and incubate at room temperature in darkness for 1 h.
13. Remove the secondary antibody working solution.
14. Rinse with PBS (×2).
15. Add counterstain working solution (see see Sect. 7.2.2, step 12) and incubate at room temperature in darkness for 2–5 min.
16. Remove the counterstain working solution.
17. Add PBS.
18. Store cells before analysis in darkness at 4°C.
19. Analyze staining results with an inverted fluorescent microscope.

References

http://www.incworld.com/_intro/ihc-methods.htm
http://www.bio.davidson.edu/Courses/genomics/method/IMF.html
http://www.biobest.co.uk/diagnostics/immunofluoresence.html
http://www.protocol-online.org/prot/Immunology/Immunofluorescence/
Amit M, Chebath J, Margulets V, Laevsky I, Miropolsky Y, Shariki K, Peri M, Blais I, Slutsky G, Revel M, Itskovitz-Eldor J (2010) Suspension culture of undifferentiated human embryonic and induced pluripotent stem cells. Stem Cell Rev 6:248–259

Chapter 8
Karyotype and Fluorescent In Situ Hybridization Analysis of Human Embryonic Stem Cell and Induced Pluripotent Stem Cell Lines

Ilana Laevsky

Abstract Cytogenetics is the science of studying human chromosomes. In cytogenetics, karyotype is the nomenclature used to describe the normal or abnormal, constitutional or acquired, chromosomal complement of an individual, tissue, or cell line. In some cases, it is necessary to use in situ hybridization methods in addition to karyotyping for karyotype clarification. Karyotype stability of human embryonic stem cells (hESCs) and induced pluripotent stem cells (iPSC) is particularly sensitive to growth conditions (e.g., growth medium, temperature and CO_2 conditions, and splitting method). In most cases, culture pressure results in chromosomal abnormalities. This chapter will discuss the common principles of karyotyping and fluorescent in situ hybridization (FISH) methods and the application of these methods to hESC and iPSC lines.

8.1 Introduction

8.1.1 Karyotype Analysis

Chromosomes, composed of protein and DNA, are distinct dense bodies located in the nuclei of cells. The 46 chromosomes in human cells consist of 22 autosomal pairs and 2 sex chromosomes: 2 X chromosomes for females, or an X and a Y chromosome for males. Genetic information is contained in the DNA of chromosomes in the form of linear sequences of bases (A, T, C, G). The total number of bases in all the chromosomes of a human cell is approximately six billion, with the number

I. Laevsky, B.A. (✉)
Stem Cell Research Center, The Rappaport Faculty of Medicine,
Technion—Israel Institute of Technology, Haifa, Israel

Department of Obstetrics and Gynecology, Rambam Health Care Campus, Haifa, Israel
e-mail: ilanala@tx.technion.ac.il

M. Amit and J. Itskovitz-Eldor, *Atlas of Human Pluripotent Stem Cells: Derivation and Culturing*, Stem Cell Biology and Regenerative Medicine,
DOI 10.1007/978-1-61779-548-0_8, © Springer Science+Business Media, LLC 2012

of individual chromosomes ranging from 50 to 250 million. A gene consists of the DNA sequence for a single trait, and ranges in size from a few thousand to two million bases. Each chromosome contains a few thousand genes.

During interphase, which is the stage of the cell cycle during which the cell spends most of its time, chromosomes are not distinctly visible under the light microscope. However, during the process of cell division, or mitosis, they become highly condensed and visible as dark distinct bodies within the nuclei of cells. Chromosomes are most easily seen in the metaphase stage of cell division. They can then be identified using certain staining techniques, called banding. Cells are cultured and then stopped in metaphase to maximize the number of suitable cells. They are spread on a slide, stained with a suitable dye, and visualized under a microscope (Fig. 8.1a). The banding techniques can be classified into two principal categories: (1) those resulting in bands distributed along the length of the whole chromosome, such as G-, Q-, and R-banding, and (2) those that stain a restricted number of specific bands and structures, such as C-banding (for the heterochromatic regions) and Ag-NOR banding (for nucleolus organizer regions).

A typical metaphase chromosome consists of two arms separated by a primary constriction or centromere (p is the short arm and q the long arm). A chromosome may be characterized by its total length, the position of its centromere, and the pattern of its bands (Fig. 8.1b). As such, a trained cytogeneticist can identify each chromosome by observing it under the microscope. Most conventional cytogenetic analyses depend on the karyotyping of G-banded metaphase chromosomes. Homologous pairs of chromosomes are arranged according to chromosome structure; each chromosome can thus be compared band for band with its homologue to detect structural changes.

A chromosome with its centromere at or near the midpoint is referred to as metacentric. The centromere of a submetacentric chromosome is somewhat distant from the midpoint. The term acrocentric refers to a chromosome with a centromere obviously off-centered (e.g., halfway between the middle and the tip of the chromosome). Certain human chromosomes contain a secondary constriction, which appears as an unstained gap, called a satellite stalk, near the tip. The chromosome segment distal to this gap appears as a satellite. Chromosome numbers 13, 14, 15, 21, and 22 have satellites.

Human karyotyping is described according to the International System for Human Cytogenetic Nomenclature (ISCN 2009). The karyotype formula begins with the total number of chromosomes in the cell, followed by the sex chromosomes (first the Xs and then the Ys). For example, the formula for a normal male is 46, XY (Fig. 8.2a) and for a normal female 46, XX (Fig. 8.2b).

An extra or a missing chromosome is designated with a + or a − sign, respectively, before the number of the chromosome. Thus, a male with trisomies (three chromosomes) for chromosomes 13 and 17 is designated as 48, XY, +13, +17 (Fig. 8.3), and a female with monosomy (one chromosome) for 12 is 45, XX, −12 (Fig. 8.4). Translocation (t) is an exchange of chromosome segments between different chromosomes. For example, the formula for a male with balanced translocation between chromosomes 18 and 19 (with breakpoint 12.2 on a long arm of chromosome 18 and 13.3 on a long arm of chromosome 19) is designated as 46, XY, t(18;19)(q12.2;q13.3) (Fig. 8.5). An addition or deletion of a chromosome segment

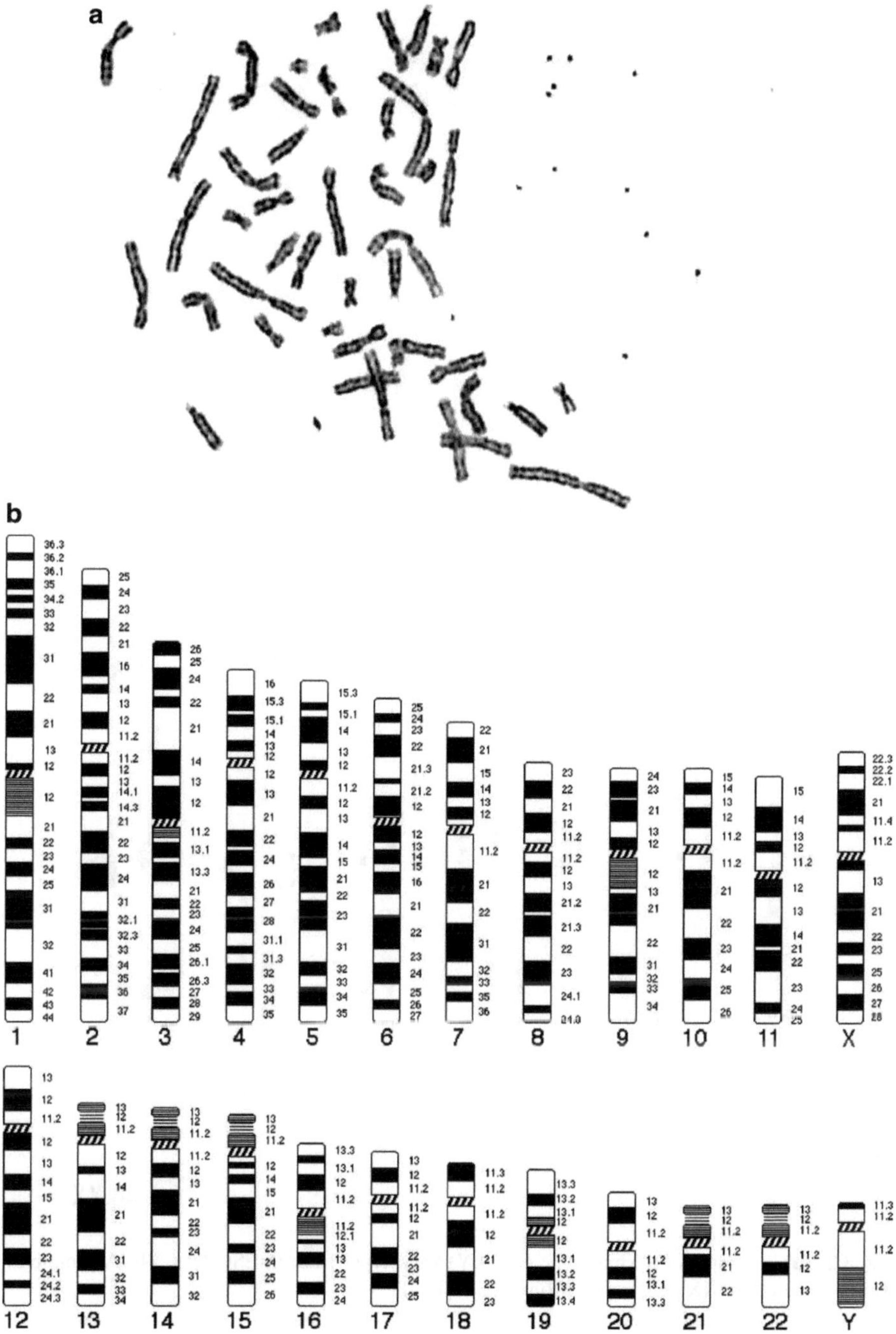

Fig. 8.1 (**a**) Metaphase chromosomes under an upright microscope (objective ×100). (**b**) Human chromosome ideograms

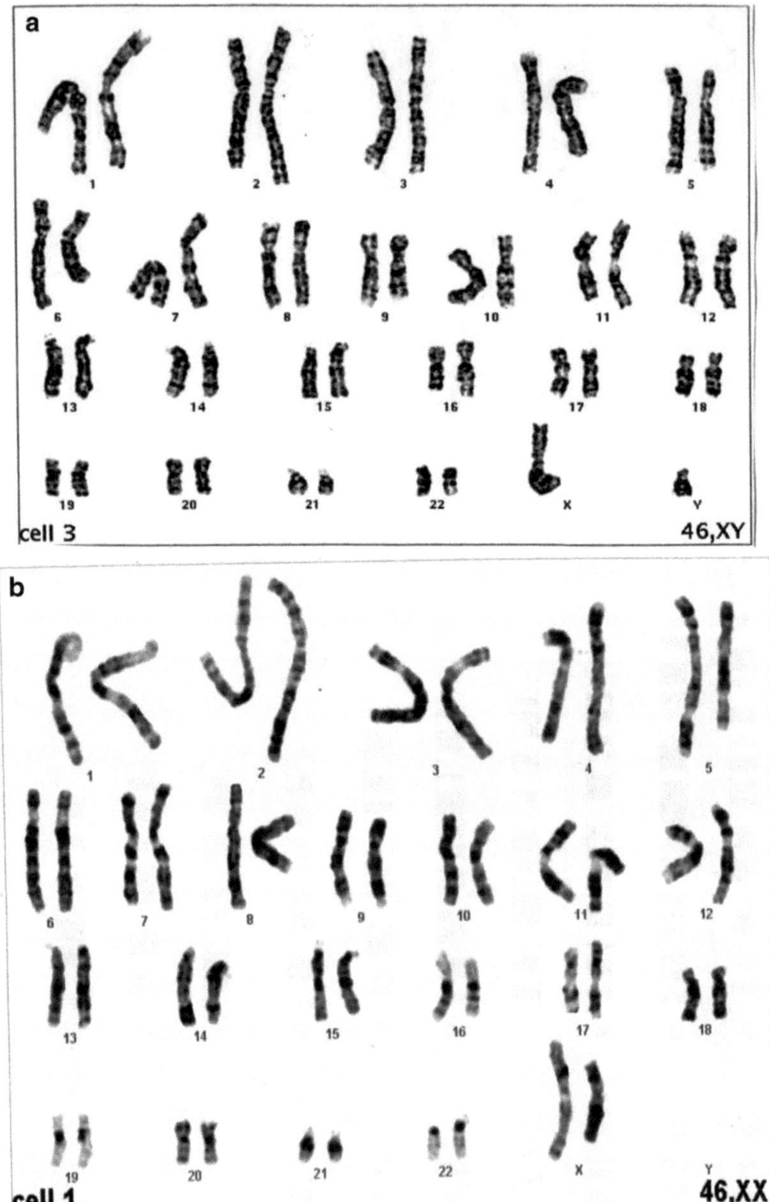

Fig. 8.2 (**a**) A sample of normal male (46, XY) karyotype (from I6 ESC line). (**b**) A sample of normal female (46, XX) karyotype (from I4 ESC line)

is denoted with a + or a − sign, respectively, after the symbol of the chromosome arm. The formula for a male with translocation between chromosomes 1 and 22 is 46, XY, +der1(t (1;22)(p10;q10)), −22 with centromere as the break point on chromosomes 1 and 22, resulting in partial trisomy of chromosome 1 (Fig. 8.6). Table 8.1 presents a partial list of symbols for chromosome nomenclature.

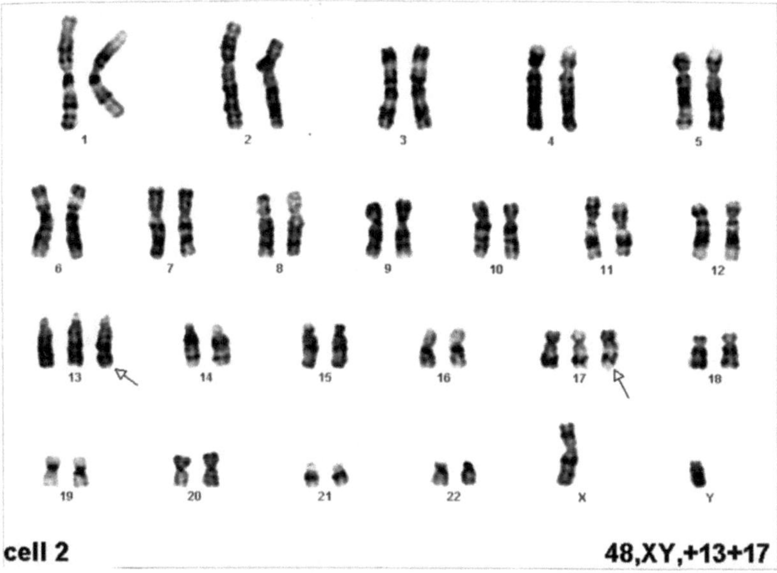

Fig. 8.3 Demonstration of abnormal karyotype with chromosome 13 and 17 trisomies (from I6 ESC line)

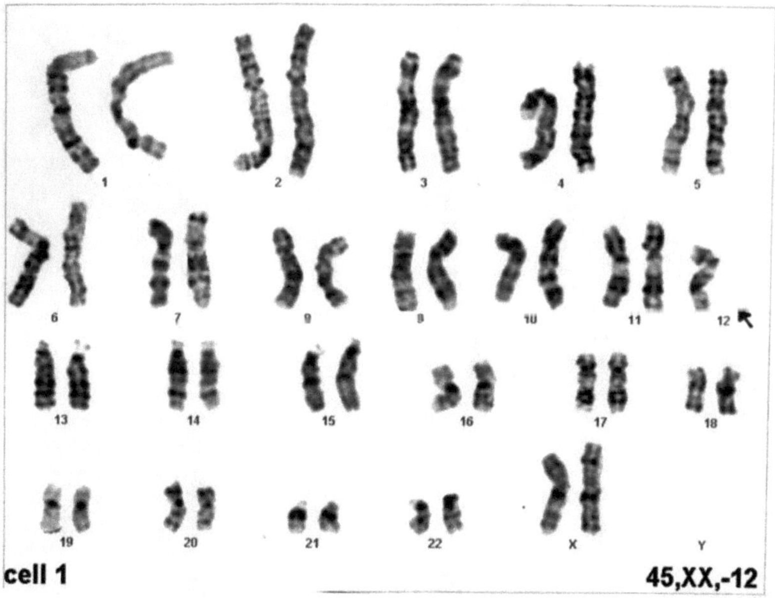

Fig. 8.4 Example of an abnormal karyotype with chromosome 12 monosomy (from I3 ESC line)

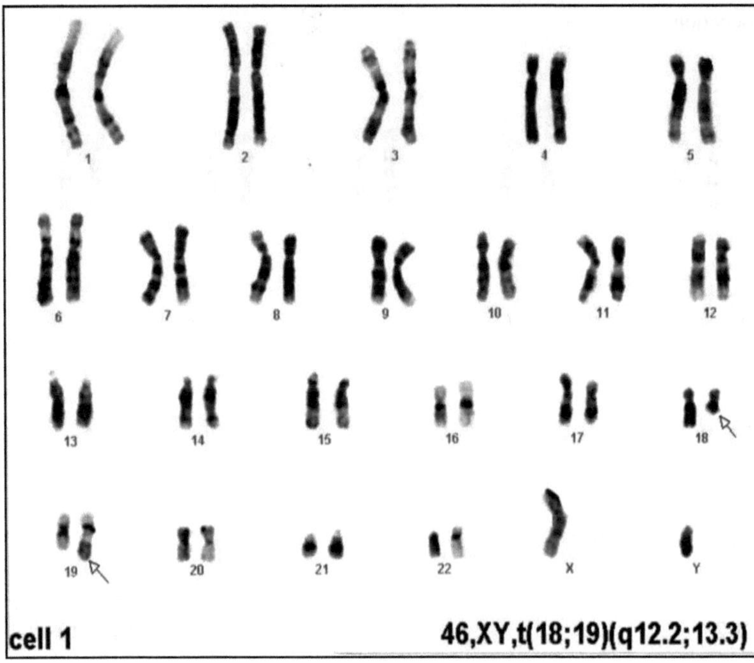

Fig. 8.5 An abnormal karyotype with balanced translocation between chromosomes 18 and 19 (from H6 ESC line)

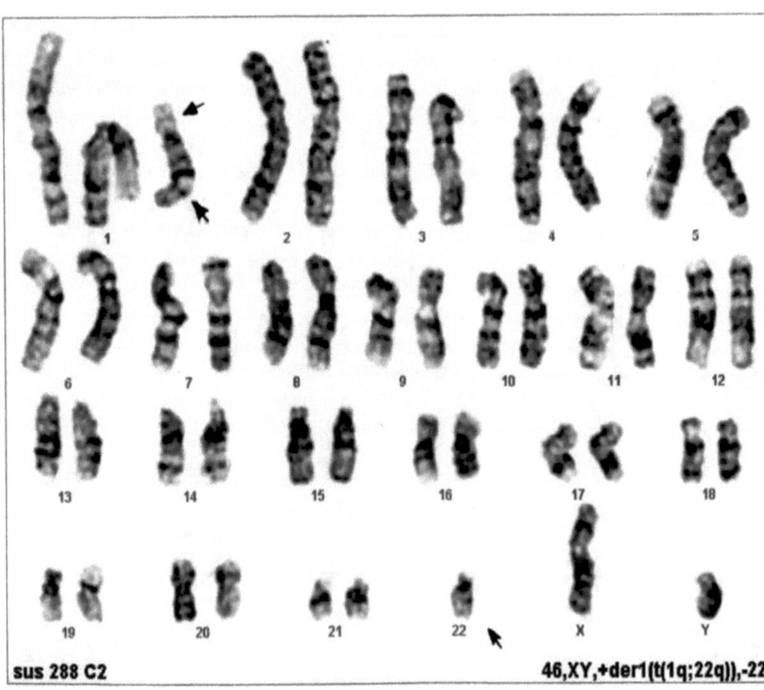

Fig. 8.6 Example of an abnormal karyotype with translocation between chromosomes 1 and 22 resulting in partial trisomy of chromosome 1 (trisomy of a long arm of chromosome 1) (from iPSC line derived from human foreskin C2)

Table 8.1 Symbols for chromosome nomenclature

A–G	The chromosome groups
1–22	The autosome numbers
X, Y	The sex chromosomes
del	Deletion
der	Derivative of chromosome
dup	Duplication
ins	Insertion
inv	Inversion
t	Translocation
p	Short arm of chromosome
q	Long arm of chromosome
s	Satellite
mar	Marker chromosome (unknown chromosome)

The pattern of light and dark bands changes with the stage of mitosis, ranging from 300 or less bands per haploid karyotype in late metaphase to more than 1,000 bands in early prophase. As mitosis progresses, the bands coalesce, accounting for the reduction of bands; resolution is lost by the fusion of sub-bands into larger bands. Most routine laboratory work is done at the 400-band level. For critical analysis, special techniques must be employed to produce 550 or more bands, known as "high-resolution analysis" (Francke 1981, 1994; Yunis 1976; Yunis et al. 1978).

8.1.2 FISH Analysis

In addition to karyotyping (classic cytogenetics), cytogeneticists use molecular cytogenetic methods based on non-isotopic in situ hybridization techniques. The most common method is fluorescent in situ hybridization (FISH), which uses sequence-specific DNA probes labeled with fluorochromes. The locations of differently tagged probes, and the relative positions of their binding sites may be visible microscopically on a single chromosome segment or DNA/chromatin fiber (Wiegant et al. 1993). There are various types of fluorescent probes: enumeration probes (for the centromere region) (Figs. 8.7a and 8.8a), locus-specific probes (for a specific chromosome segment or gene) (Figs. 8.7b and 8.8a), whole chromosome painting probes (Fig. 8.8b), all telomere probes, all NOR probes, and variations of these types.

Unlike classic karyotyping methods, which are only applicable to mitotic metaphase chromosomes, the FISH technique can be applied to all stages of the cell cycle. As such, FISH techniques have enhanced the capability of identifying chromosome segments, of correlating chromosome structures with gene locations, of revealing cryptic abnormalities that are undetectable using standard banding techniques, and of analyzing and describing complex rearrangements (Trask 1991).

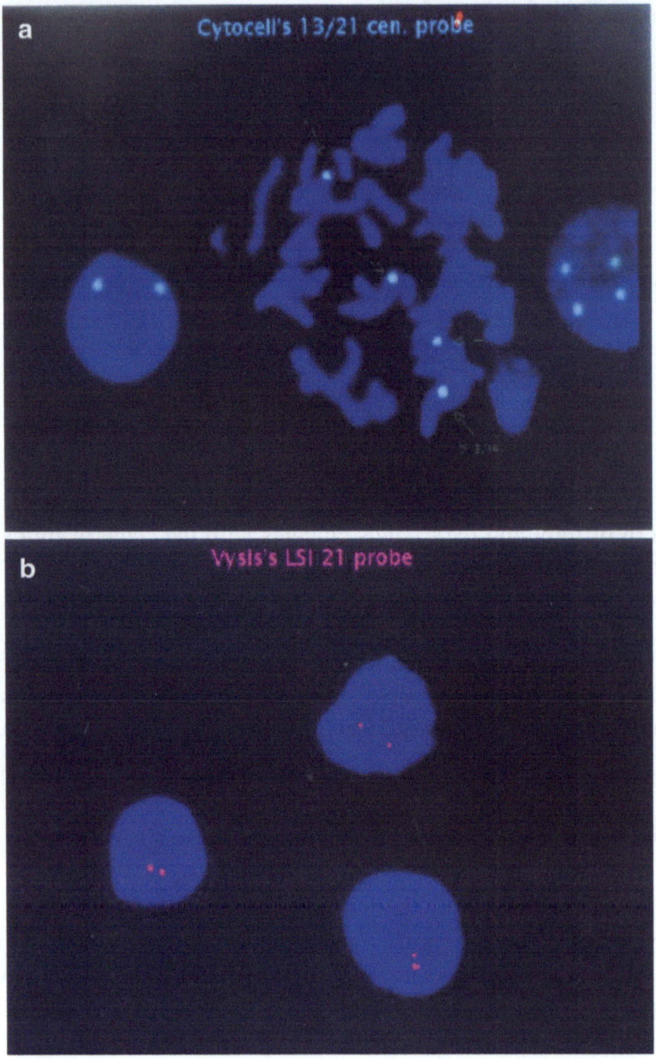

Fig. 8.7 (**a**) FISH with enumeration probe for centromeres of chromosomes 13 and 21. There are two copies of chromosome 13 and two copies of chromosome 21 (marked by *arrow*) in a normal cell, for a total of 4 signals in a normal cell. (**b**) FISH with locus-specific probe to the critical region of Down syndrome on chromosome 21. The presence of two signals indicates the cell is normal (3 signals per cell indicates Down syndrome). All illustrated cells are normal

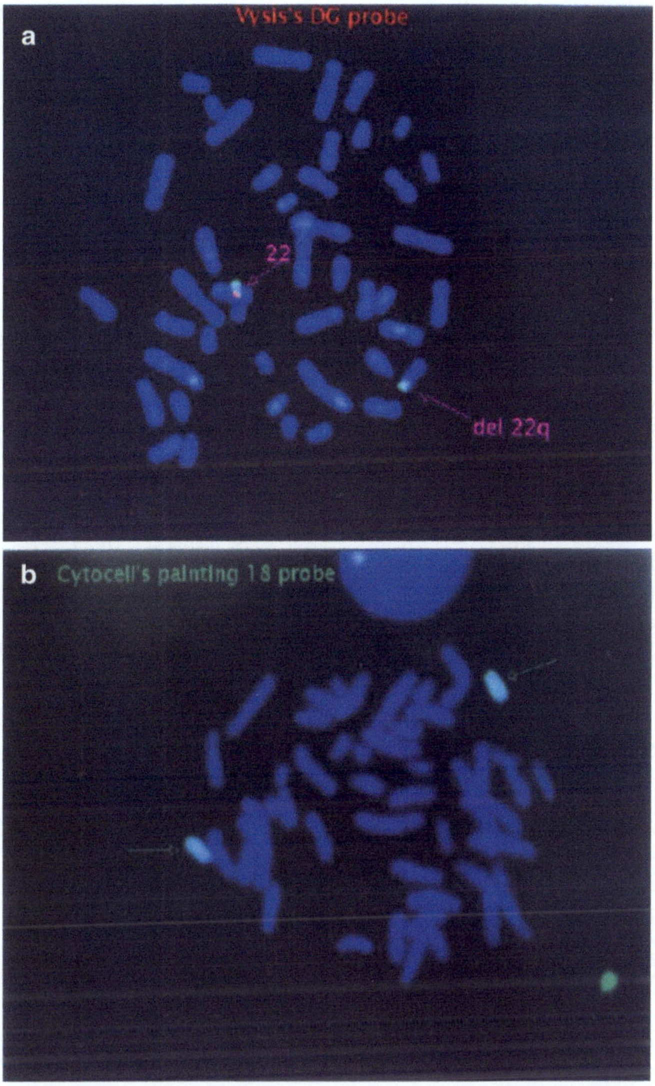

Fig. 8.8 (**a**) FISH with the probe to the gene of DiGeorge Syndrome on chromosome 22. The probe combined from the enumeration probe to the centromere of chromosome 22 (*green* signal) and the locus-specific probe to the DiGeorge gene on the 22q arm (*red* signal). A cell with two *green* signals and two *red* signals is a normal cell. A cell with two *green* signals (two chromosomes 22) and one *red* signal (deletion of the DiGeorge gene region) represents abnormalcy or DiGeorge Syndrome, as shown in the example. (**b**) An example of normal FISH with the whole chromosome painting a probe to chromosome 18

If FISH further clarifies the karyotype and enables detection of an abnormality, then the karyotype may be rewritten to reflect this new FISH information.

8.2 Materials for Harvesting Cells for Karyotyping and FISH Analysis

8.2.1 Reagents

1. Chromosome Resolution Additive (CRA) (GGS-Genial Genetic Solutions Ltd., Zotal, cat GGS-JL-003A)—DNA-intercalating reagent for inhibition of chromosome condensation.
2. Colcemid (Gibco, CA; KaryoMAX Colcemid liquid, cat #15212-012)—for inhibition of spindle formation.
3. Glacial acetic acid (ACS grade) (MERCK, Germany).
4. Absolute methanol (ACS grade) (MERCK, Germany).
5. KCl (MERCK, Germany).
6. Na citrate (MERCK, Germany).
7. Hanks' salt solution (GIBCO, CA; Biological Industries Ltd).
8. Trypsin–EDTA (×10) (GIBCO, CA).

8.2.2 Solutions

8.2.2.1 CRA Working Solution

1:100 (CRA: Hanks salt solution)—store for 1 week at 4–8°C.

8.2.2.2 Hypotonic Solution

KCl 0.56 g + Na citrate 0.5 g/200 ml distilled water.

8.2.2.3 Fixer Solution

1:3 (acetic acid: methanol)—store for only 1 day at −20°C.

8.3 Procedure of Harvesting Cells for Karyotyping and FISH Analysis

The following protocol was formulated for use with one well in a 6-well plate (2 ml of medium, 10 cm^2). For other conditions, the quantities should be adapted accordingly. Harvesting cells 3–4 days post-splitting is highly recommended if collagenase

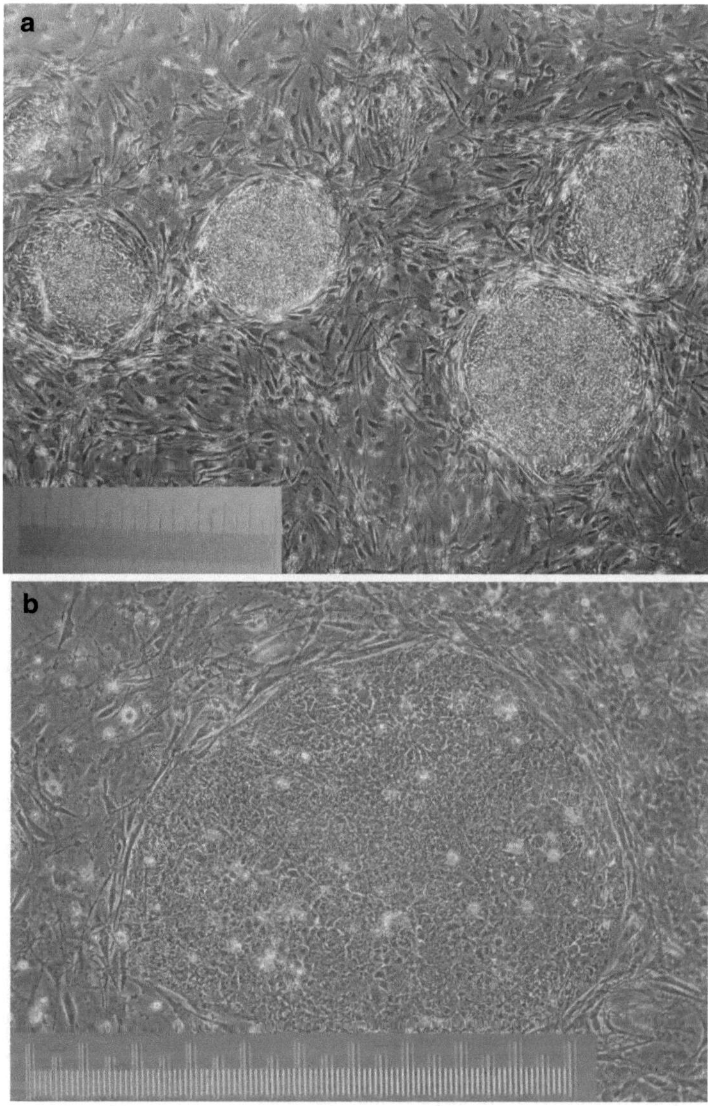

Fig. 8.9 (**a**) ESC culture ready for karyotyping (objective ×5) (from H9.2 ESC line, 3 days post-collagenase splitting). (Bar = 1 mm). (**b**) ESC colony ready for karyotyping (objective ×10) (from H9.2 ESC line, 3 days post-collagenase splitting). (Bar = 1 mm)

enzyme is used, and 4–6 days post-splitting if trypsin is used. The morphology of colonies of a culture ready for harvest is illustrated in Fig. 8.9. The use of at least one well is advised.

When the culture is ready, harvest the cells as follows:

1. Add 30 μl of sterile CRA working solution to 2 ml of fresh ESC medium.
2. Incubate for 1 h at 37°C (in the incubator).

3. Add 150 µl of sterile colcemid to the well.
4. Incubate for 30 min at 37°C (in the incubator).
5. Remove the medium.
6. Add 0.5 ml of trypsin per well.
7. Incubate for 5–10 min at 37°C for complete tissue breaking.
8. Add 1 ml of medium per well and collect the cells.
9. Centrifuge for 5 min at 1,200–1,300 rpm (using a 15-ml conical centrifugation tube).
10. Aspirate the upper liquid leaving the pellet.
11. Add 2 ml of hypotonic solution per tube, pre-warmed to 37°C.
12. Incubate for 30 min at 37°C (in a water bath).
13. Add 1 ml of fixer solution to the hypotonic solution, drop by drop, and rock gently by hand.
14. Incubate for 5 min at room temperature.
15. Centrifuge for 5 min at 1,200–1,300 rpm.
16. Aspirate the upper liquid.
17. Add 2 ml of fixer solution per tube.
18. Centrifuge for 5 min at 1,200–1,300 rpm.
19. Aspirate the upper liquid.
20. Repeat steps 17–19 twice.

Store at −20°C. The pellet can be stored for several months before use. The chromosome analysis should be performed by a specialist in a cytogenetic unit. A pellet is dropped onto the slides, stained using trypsin and Giemsa, analyzed under a microscope, and then karyotyped.

References

Francke U (1981) High-resolution ideograms of trypsin-Giemsa banded human chromosomes. Cytogenet Cell Genet 31:24–32

Francke U (1994) Digitized and differentially shaded human chromosome ideograms for genomic applications. Cytogenet Cell Genet 65:206–218

Shaffer LG, Slovak ML, Campbell LJ (eds) (2009) ISCN 2009: an international system for human cytogenetic nomenclature. S. Karger, Basel, Switzerland

Trask BJ (1991) Fluorescence in situ hybridization: applications in cytogenetics and gene mapping. Trends Genet 7:149–154

Wiegant J, Wiestmeijer CC, Hoovers JM, Schuuring E, d'Azzo A, Vrolijk J, Tanke HJ, Raap AK (1993) Multiple and sensitive fluorescence in situ hybridization with rhodamine-, fluorescein-, and coumarin-labeled DNAs. Cytogenet Cell Genet 63:73–76

Yunis JJ (1976) High resolution of human chromosomes. Science 191:268–1270

Yunis JJ, Sawyer JR, Ball DW (1978) The characterization of high-resolution G-banded chromosomes of man. Chromosoma 67:293–307

Chapter 9
Method for the Derivation of Induced Pluripotent Stem Cells from Human Hair Follicle Keratinocytes

Atara Novak

Abstract Induced pluripotent stem cells (iPSCs) are pluripotent stem cells that are artificially derived from non-pluripotent cells by inducing "forced" expression of specific genes. Since first reported in 2006, iPSC technology has attracted great interest for its potential application in regenerative medicine and cell therapy. Cell types that require harvesting by medical intervention, such as skin fibroblasts and blood, have been the most common sources for deriving human iPSCs. The need to identify an alternative cell source that can be easily obtained from patients without an invasive procedure is well recognized. Here, we describe a simple method for generating human iPSCs from hair keratinocytes derived from the outer root sheath of plucked hair follicles, using a single polycistronic excisable lentiviral vector—the "STEMCCA" cassette.

9.1 Introduction

Induced pluripotent stem cells (iPSCs) are human somatic cells that have been genetically reprogrammed into a pluripotent state similar to that of human embryonic stem cells (hESCs). They are obtained from various cell types of both mice and humans by the exogenous expression of different combinations of Oct4, Sox2, Klf4, c-Myc, Nanog, and Lin28 (Takahashi et al. 2007; Yu et al. 2007). iPSCs have

A. Novak, M.Sc. (✉)
Stem Cell Research Center, The Rappaport Faculty of Medicine,
Technion—Israel Institute of Technology, Haifa, Israel

Department of Obstetrics and Gynecology, Rambam Health Care Campus, Haifa, Israel
e-mail: atarap@tx.technion.ac.il

M. Amit and J. Itskovitz-Eldor, *Atlas of Human Pluripotent Stem Cells: Derivation and Culturing*, Stem Cell Biology and Regenerative Medicine,
DOI 10.1007/978-1-61779-548-0_9, © Springer Science+Business Media, LLC 2012

generated great interest in regenerative medicine due to their capability of generating patient- and disease-specific pluripotent stem cells. This technique can provide effective platforms for the discovery of new drugs, and elucidate disease mechanisms that may ultimately lead to the development of cell and tissue replacement therapies (Kiskinis and Eggan 2010). Various types of somatic cells that provide sources for generating human iPSCs, fibroblasts being the most common of them (Huangfu et al. 2008; Lowry et al. 2008; Park et al. 2008; Soldner et al. 2009; Takahashi et al. 2007), are isolated from tissues harvested via surgical intervention.

Aasen et al. were the first to report the efficient reprogramming of human keratinocytes derived from skin biopsies (Aasen et al. 2008). Skin fibroblasts are isolated through an invasive surgical procedure. Skin biopsies are not always possible due to blood coagulation disorders, skin diseases, or mental disorders. Moreover, young children cannot undergo this procedure and some adults may refuse it for religious or cultural reasons (Novak et al. 2010).

Aasen et al. also demonstrated that keratinocytes derived from human plucked hair can be reprogrammed into iPSCs (Aasen et al. 2008). Reprogramming is achieved by culturing the hair in MEF-conditioned hESC medium on a Matrigel-coated dish, using the pMSCV retrovirus vectors expressing Yamanaka's four reprogramming factors. As an effortlessly obtained biological material, readily available from all individuals, plucked hair has significant advantages over skin fibroblasts as a source for the generation of human iPSCs. Plucked hair follicles have thus become a very practical and convenient biopsy material for the study of genetic disorders, as well as for diagnostic purposes (Limat and Noser 1986).

A hair follicle is composed of concentric layers containing the inner sections of the hair shafts, the surrounding inner root sheath (IRS), and the outer root sheath (ORS), which is the outermost layer of the hair follicle (Rogers 2004; Schneider et al. 2009). Sommer and his colleagues recently reported on the use of a single excisable lentiviral "stem cell cassette" (STEMCCA) encoding the four reprogramming factors—Oct4, Sox2, Klf4, and c-Myc—in a single polycistronic vector flanked by loxP sites (Sommer et al. 2009, 2010).

In this chapter, we provide a detailed protocol for the reprogramming of human keratinocytes derived from the ORS of human plucked hair follicles, using the humanized version of the single lentiviral "STEMCCA" vector. The derived iPSCs were further characterized for their lineage-specific differentiation potential in vitro through spontaneous EB formation, and in vivo using the teratoma assay. Moreover, we found this reprogramming protocol to be efficient due to the fact that the hair keratinocytes are sensitive and, unlike the fibroblasts, cannot be grown in hESC conditions (Novak et al. 2010). Therefore, all resultant colonies comprise true and stable iPSCs that can be easily observed, isolated, and further expanded. Our data suggest that human hair follicle keratinocytes, reprogrammed by the STEMCCA cassette-generated iPSCs, represent a superior source for modeling human diseases and clinical applications.

9.2 Materials

9.2.1 NIH-3T3/293T Cells

9.2.1.1 Culture Medium

90% Dulbecco's modified Eagle's medium (DMEM) (GIBCO, 14190-094), 10% fetal bovine serum (FBS) (HyClone, SV30160.03), 2 mM L-glutamine (Biological Industries Ltd., 03-020-1B), 100 U/ml penicillin, and 100 μg/ml streptomycin (Biological Industries Ltd., 03-031-1B).

9.2.1.2 Freezing Medium

40% DMEM, 40% FBS, and 20% dimethyl sulfoxide (DMSO) (Sigma, D-2650).

9.2.1.3 Splitting

Trypsin/EDTA (Invitrogen Corporation, type IV, 17104019).

9.2.1.4 Washing

Phosphate-buffered saline (PBS) with Ca^- and Mg^- (GIBCO, 14190-094).

9.2.1.5 (For NIH-3T3) Mitomycin C

8 μg/ml mitomycin C (Sigma, M-4287) diluted in DMEM.

9.2.2 Keratinocyte Derivation from Plucked Hair Follicles

9.2.2.1 Follicle Washing Medium

DMEM containing 25 mM HEPES (Biological Industries Ltd, 03-025-1B), 2 mM L-glutamine, 400 U/ml penicillin, and 400 μg/ml streptomycin (PS).

9.2.2.2 Culture Medium (Green Medium)

60% DMEM, 30% DMEM F-12, 10% FBS, 1 mM sodium pyruvate (Sigma, P-5280), 2 mM L-glutamine, 5 μg insulin (Sigma, I-9278), 0.5 μg/ml hydrocortisone (Sigma, H-0888), 0.2 nM adenine (Sigma, A-2786), 2 nM triiodothyronine (T3) (Sigma, T-2877), 10 ng/ml epidermal growth factor (EGF) (R&D Systems, 236-EG), transferrin (GIBCO, 03-0124SA), 100 U/ml penicillin, and 100 μg/ml streptomycin.

9.2.2.3 Splitting

3T3 Removal

0.02% EDTA (Promega, V4231) in PBS.

Keratinocyte Splitting

0.1% Trypsin and 0.02% EDTA (Invitrogen Corporation, type IV, 17104019) in PBS.

9.2.2.4 Washing

PBS with Ca^- and Mg^-.

9.3 Methods

9.3.1 NIH-3T3 and 293T Culture Methods

9.3.1.1 NIH-3T3/293T Splitting

1. Remove the culture medium (see Sect. 9.2.1.1).
2. Wash with 5 ml of PBS (for T75 flask) to remove all traces of serum that contain trypsin inhibitor.
3. Add 3 ml of trypsin/EDTA.
4. Incubate for 5 min at 37°C.
5. Tap the side of the flask to loosen the cells. Add 5 ml of culture medium (see Sect. 9.2.1.1) to neutralize the trypsin.

6. Transfer the cell suspension into a conical tube.
7. Centrifuge for 5 min at 250×*g*.
8. Remove the suspension, re-suspend in 3 ml of culture medium (see Sect. 9.2.1.1), and pipette in order to fracture the pellet.
9. Distribute the cell suspension to the desired number of culture flasks. The recommended ratio is 1:6–1:8.
10. Add the culture medium (see Sect. 9.2.1.1) to a final volume of 15 ml.

Do not allow the cells to be more than 80% confluent.

9.3.1.2 NIH-3T3/293T Freezing

1. Wash with 5 ml of PBS (for T75 flask) to remove all traces of serum that contain trypsin inhibitor.
2. Add 3 ml of trypsin/EDTA and cover the entire culture flask surface.
3. Incubate for 5 min.
4. Tap the side of the flask to loosen the cells. Add 5 ml of culture medium (see Sect. 9.2.1.1) to neutralize the trypsin.
5. Transfer the cell suspension into a conical tube.
6. Centrifuge for 5 min at 250×*g*.
7. Remove the suspension, re-suspend in 1.5 ml of culture medium (see Sect. 9.2.1.1), and pipette in order to fracture the pellet.
8. Add an equivalent volume of freezing medium drop by drop (see Sect. 9.2.1.1) and mix gently. Adding the freezing medium drop by drop is crucial for cell recovery.
9. Place 1 ml into two 1-ml cryogenic vials.
10. Freeze the vials overnight at −80°C in a freezing box (Nalgene freezing box, C.N. 5100–0001).
11. Transfer the vials into a liquid nitrogen container.

9.3.1.3 NIH-3T3/293T Thawing

1. Remove the vial from the liquid nitrogen and thaw briefly in a 37°C water bath.
2. When a small pellet of frozen cells remains, clean the vial using 70% ethanol.
3. Pipette the contents of the vial once and transfer the cells into a conical tube.
4. Add 2 ml of culture medium drop by drop (see Sect. 9.2.1.1). Adding the medium drop by drop is crucial for cell recovery.
5. Centrifuge for 5 min at 250×*g*.
6. Re-suspend the pellet in the culture medium (see Sect. 9.2.1.1).
7. Transfer the cell suspension into culture flasks and add 15 ml of culture medium (see Sect. 9.2.1.1).

9.3.1.4 Preparation of NIH-3T3 Covered Plates

1. Add 8 μg/ml of mitomycin C (see Sect. 9.2.1.3) into the culture flask and incubate for 2 h.
2. Wash four times with 10 ml PBS.
3. Add 3 ml of trypsin/EDTA and cover the entire culture flask surface (T75).
4. Incubate for 5 min.
5. Tap the side of the flask to loosen the cells. Add 4 ml of NIH-3T3 culture medium (see Sect. 9.2.1.1) to neutralize trypsin.
6. Transfer the cell suspension into a conical tube.
7. Centrifuge for 5 min at $250 \times g$.
8. Remove the suspension, re-suspend in 10 ml of culture medium (see Sect. 9.2.1.1), and pipette in order to fracture the pellet.
9. Count the cells and re-suspend in the desired medium volume (see Sect. 9.2.1.1).
10. Seed 2×10^5 inactivated NIH-3T3 cells per well in 6-well plates (10 cm^2) per 2 ml (2×10^4 cells/cm^2). Let set for at least 8 h before plating the keratinocytes.

9.3.2 Keratinocyte Culture Methods

9.3.2.1 Derivation of Keratinocytes from Plucked Hair Follicles

Human hair keratinocytes can be derived as previously described (Limat and Noser 1986).

1. Pluck ten hair follicles with the visible ORS from the scalp.
2. Cut off the bulk of the hair follicles.
3. Immerse the follicles into a 10-cm petri dish with hair washing medium (see Sect. 9.2.1.1) for 30 min in 37°C.
4. Remove the hair washing medium and add 0.1% trypsin and 0.02% EDTA.
5. Incubate for 15–30 min at 37°C until cells are visibly dissociated from the follicles (see Fig. 9.1a, b).
6. Pipette the follicles vigorously with Green medium (see Sect. 9.2.1.2) to obtain a single-cell suspension.
7. Centrifuge the dissociated keratinocytes for 10 min at $200 \times g$ and seed in two wells of a 6-well plate on NIH-3T3 feeder layer (2×10^4 3T3 cells/cm^2) with Green medium (see Fig. 9.1c, d).

9.3.2.2 Keratinocyte Splitting

1. Aspirate the culture medium and wash the well once with 2 ml of PBS (one well of a 6-well plate).
2. Add 1 ml of 0.02% EDTA to remove the NIH-3T3 feeder cells.
3. Incubate for 5 min at 37°C.

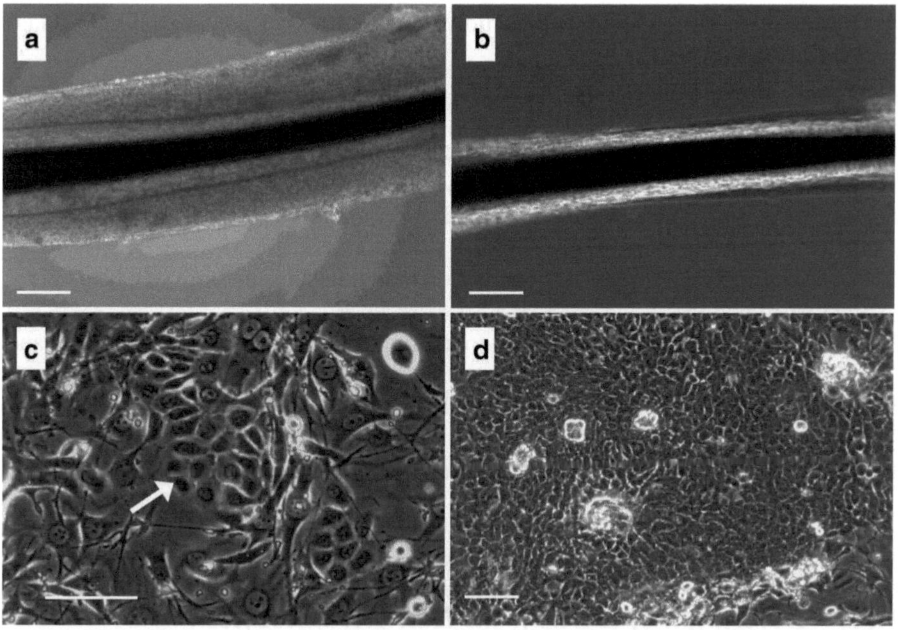

Fig. 9.1 Derivation of human hair follicle keratinocytes. (**a**) A bulk of intact plucked hair follicles. (**b**) An intact plucked hair follicle following enzymatic removal of the cells. (**c**) Hair keratinocytes (at *arrow*), which were isolated from the plucked hair follicles seeded on inactivated NIH-3T3 feeder cells (around the keratinocytes), appeared as small colonies 2 days after passaging and (**d**) as large colonies 7 days after passaging

4. Wash the culture twice with 2 ml of PBS.
5. Add 1 ml 0.1% of trypsin and 0.02% of EDTA to detach the keratinocyte cells into single cells.
6. Incubate for 10–15 min at 37°C.
7. Centrifuge the dissociated keratinocytes for 5 min at $200 \times g$.
8. Seed 30,000 keratinocytes on inactivated NIH-3T3 feeder layer (2×10^4 3T3 cells/cm^2).

9.3.3 Preparation of the STEMCCA Virus for Infection

The STEMCCA vector is a lentivirus containing the four factors: OCT4, SOX2, KLF4, and C-MYC, generated by Dr. Gustavo Mostoslavsky at Boston University School of Medicine.

1. Day 1: Seed 2 million 293T cells in culture in a 10-cm petri dish with 7 ml of growth medium (see Sect. 9.2.1.1).
2. Day 2: Transfect to 293T cells with a total DNA quantity of 15 μg per 10-cm petri dish, using JET reagent (Tamar Ltd.) according to the manufacturer's

instructions. The lentiviral plasmid ratio is 20:1:1:1:2—STEMCCA: Gag-Pol:REV:TAT:VSVG.

3. Day 3: Replace the growth medium with 4.5 ml of Green medium (see Sect. 9.2.2.2).
4. Day 4: Collect the Green medium containing the virus from the 293T cells and replace it with 4.5 ml of fresh Green medium.
5. Filter the infection medium (Green medium containing the virus) with a 0.45-μm filter.
6. Add 2 μg/ml of polybrene to the infection medium. This medium will be used for the first infection.
7. Day 5: Collect the Green medium containing the virus from the 293T cells and eliminate the 293T cells.
8. Filter the infection medium (Green medium containing the virus) with a 0.45-μm filter.
9. Add 2 μg/ml of polybrene to the infection medium. This medium will be used for the second infection.

9.3.4 Derivation of iPSCs from Hair Keratinocytes

1. On day 1, seed 30,000 keratinocytes in one well of a 6-well plate on inactivated NIH-3T3 feeder layer (see Sect. 9.3.2).
2. On day 4, after seeding the keratinocytes, aspirate the culture medium and wash the well once with 2 ml of PBS (for one well of a 6-well plate).
3. Add 1 ml of 0.02% EDTA to remove the NIH-3T3 feeder cells.
4. Incubate for 5 min at 37°C.
5. Wash the culture twice with 2 ml of PBS.
6. First infection. Add 4.5 ml of filtrated infection medium containing the STEMCCA lentivirus and polybrene (see Sect. 9.3.3) to the keratinocyte culture (free of feeder) in one well of a 6-well plate.
7. Centrifuge the infection for 50 min with $500 \times g$ at 32°C.
8. Remove the infection medium.
9. Wash the cells with PBS.
10. Add 2×10^5 inactivated NIH-3T3 cells with 2 ml of Green medium (see Sect. 9.2.2.2) to one well of a 6-well plate containing the infected keratinocytes.
11. Second infection: Repeat the infection protocol after 24 h.
12. At day 4 postinfection, aspirate the culture medium and wash the well once with 2 ml of PBS (one well of a 6-well plate).
13. Add 1 ml of 0.02% EDTA to remove the NIH-3T3 feeder cells.
14. Incubate for 5 min at 37°C.
15. Wash the culture twice with 2 ml of PBS.
16. Add 1 ml 0.1% of trypsin and 0.02% EDTA to detach the keratinocyte cells into single cells.
17. Incubate for 10–15 min at 37°C.

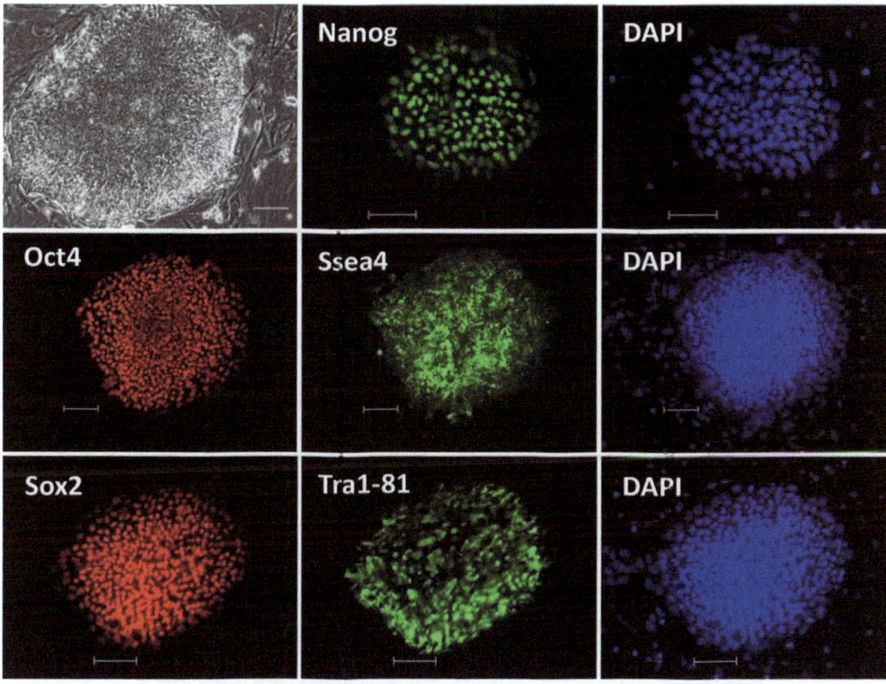

Fig. 9.2 Pluripotency of iPSCs derived from hair keratinocytes. Morphology of an iPSC colony derived from hair keratinocytes, clone KTR13 at passage 27, and immunostaining of the typical hESC markers; Nanog, Oct4, Tra1-81, SSEA4, and Sox2 shown for the iPSC clone KTR13 derived from hair keratinocytes at passage 20. Nuclei are stained with DAPI (*blue*). Scale bar represents 100 μm

18. Centrifuge the dissociated keratinocytes for 5 min at 200×*g*.
19. Seed the dissociated keratinocytes on three wells of a 6-well plate covered with 4×10^5 inactivated MEFs (see Sect. 3.3.1.6 in Chap. 3) with 2 ml of Green medium (see Sect. 9.2.2.2) per well.
20. On the next day, replace the Green medium with 4 ml of hESC medium (see Chap. 3, serum-free medium, Sect. 3.2.2.1) containing 8 ng/ml bFGF per one well of a 6-well plate.
21. Replace the medium every other day.

iPSC colonies emerge about 25 days after keratinocyte seeding on the MEF feeder; they are true iPSCs according to immunostaining of typical hESC markers: Nanog, Oct4, Ssea4, Sox2, and Tra1-60 (see Fig. 9.2).

The pluripotency of the iPSCs has been further characterized by the differentiation of the cells; in vitro into embryonic bodies (EBs) and in vivo into teratomas. The differentiated cells contained derivatives of all three embryonic germ layers; ectoderm, mesoderm, and endoderm (see Fig. 9.3).

For the maintenance of iPSCs, see Sect. 3.2.2 in Chap. 3.

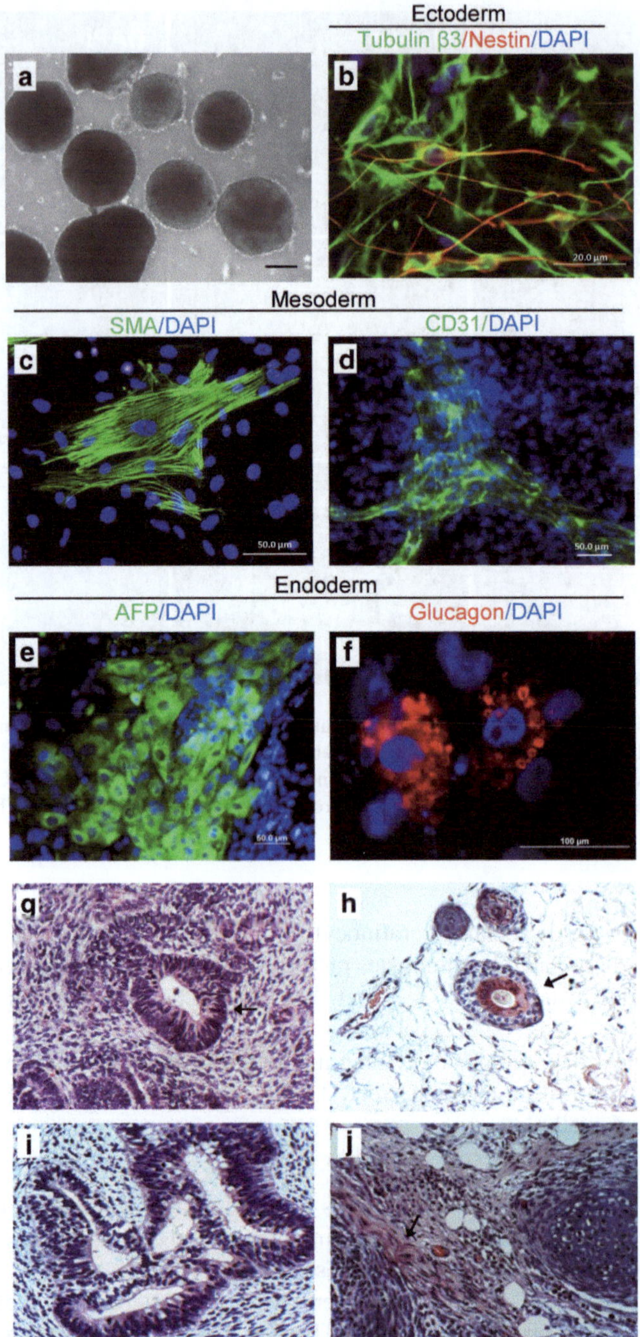

Fig. 9.3 Differentiation of HFKT-iPSCs. (**a**) Five-day-old EBs generated from the iPSCs derived from hair keratinocytes, KTR13 clone. (**b–f**) Immunostaining of 21-day-old EBs derived from the KTR13 clone revealed expression of ectodermal (tubulinβ3, Nestin) (**b**), mesodermal (SMA—**c**,

References

Aasen T, Raya A, Barrero MJ, Garreta E, Consiglio A, Gonzalez F, Vassena R, Bilic J, Pekarik V, Tiscornia G et al (2008) Efficient and rapid generation of induced pluripotent stem cells from human keratinocytes. Nat Biotechnol 26:1276–1284

Huangfu D, Osafune K, Maehr R, Guo W, Eijkelenboom A, Chen S, Muhlestein W, Melton DA (2008) Induction of pluripotent stem cells from primary human fibroblasts with only Oct4 and Sox2. Nat Biotechnol 26:1269–1275

Kiskinis E, Eggan K (2010) Progress toward the clinical application of patient-specific pluripotent stem cells. J Clin Invest 120:51–59

Limat A, Noser FK (1986) Serial cultivation of single keratinocytes from the outer root sheath of human scalp hair follicles. J Invest Dermatol 87:485–488

Lowry WE, Richter L, Yachechko R, Pyle AD, Tchieu J, Sridharan R, Clark AT, Plath K (2008) Generation of human induced pluripotent stem cells from dermal fibroblasts. Proc Natl Acad Sci U S A 105:2883–2888

Novak A, Shtrichman R, Germanguz I, Segev H, Zeevi-Levin N, Fishman B, Mandel YE, Barad L, Domev H, Kotton D et al (2010) Enhanced reprogramming and cardiac differentiation of human keratinocytes derived from plucked hair follicles, using a single excisable lentivirus. Cell Reprogram 12:665–678

Park IH, Lerou PH, Zhao R, Huo H, Daley GQ (2008) Generation of human-induced pluripotent stem cells. Nat Protoc 3:1180–1186

Rogers GE (2004) Hair follicle differentiation and regulation. Int J Dev Biol 48:163–170

Schneider MR, Schmidt-Ullrich R, Paus R (2009) The hair follicle as a dynamic miniorgan. Curr Biol 19:R132–142

Soldner F, Hockemeyer D, Beard C, Gao Q, Bell GW, Cook EG, Hargus G, Blak A, Cooper O, Mitalipova M et al (2009) Parkinson's disease patient-derived induced pluripotent stem cells free of viral reprogramming factors. Cell 136:964–977

Sommer CA, Sommer AG, Longmire TA, Christodoulou C, Thomas DD, Gostissa M, Alt FW, Murphy GJ, Kotton DN, Mostoslavsky G (2010) Excision of reprogramming transgenes improves the differentiation potential of iPS cells generated with a single excisable vector. Stem Cells 28:64–74

Sommer CA, Stadtfeld M, Murphy GJ, Hochedlinger K, Kotton DN, Mostoslavsky G (2009) Induced pluripotent stem cell generation using a single lentiviral stem cell cassette. Stem Cells 27:543–549

Takahashi K, Tanabe K, Ohnuki M, Narita M, Ichisaka T, Tomoda K, Yamanaka S (2007) Induction of pluripotent stem cells from adult human fibroblasts by defined factors. Cell 131:861–872

Yu J, Vodyanik MA, Smuga-Otto K, Antosiewicz-Bourget J, Frane JL, Tian S, Nie J, Jonsdottir GA, Ruotti V, Stewart R et al (2007) Induced pluripotent stem cell lines derived from human somatic cells. Science 318:1917–1920

Fig. 9.3 (continued) CD31—**d**) (**d**), and endodermal (AFP—**e**, Glucagon—**f**) marker proteins. Nuclei are stained with DAPI (*blue*). Scale bar represents 100 μm for Fig. (**a**, **f**); 50 μm for Figs (**c**, **d**, **e**); and 20 μm for Fig. (**b**, **g**–**j**). Histological analysis of a representative teratoma obtained from in vitro-differentiated iPSCs derived from hair keratinocytes, KTR13 clone. The formed teratomas contained derivatives of all three embryonic germ layers (ectoderm, mesoderm, and endoderm). (**g**) Neuronal tissue and (**h**) hair follicles represent ectodermal lineage. (**i**) Endodermal epithelium with prominent mucus-producing cells representing endoderm formation. (**j**) Muscle tissues (at *arrow*), cartilage (right of image), and adipose cells, indicating mesoderm formation. Scale bar represents 50 μm

About the Authors

Michal Amit, Ph.D., is a senior scientist at the Sohnis and Forman Families Stem Cell Center, Faculty of Medicine, Technion—Israel Institute of Technology, Haifa. She obtained her Ph.D. degree in Medical Sciences from the Technion. In 1998, she trained at the University of Wisconsin, Madison, in embryonic stem cell (ESC) culture and derivation methods, at the time of the derivation of the first human ESC lines. At the Faculty of Medicine of the Technion she derived the first human embryonic stem cell lines in Israel, as part of her Ph.D. degree. Under grants funded by the National Institute of Health, Michal Amit was responsible for the Technion's ESC bank, which distributed hESCs to laboratories worldwide, and for establishing and running international courses on culturing ESCs. In recent years, she directed the Stem Cell Infrastructure Center of the Technion and has developed culture techniques for ESCs using defined and animal-free conditions. She currently devotes her research efforts toward developing methods for the mass production of pluripotent stem cells, to promote research, clinical and industrial applications.

Joseph Itskovitz-Eldor, M.D., D.Sc., is director of the Department of Obstetrics and Gynecology at Rambam Health Care Campus and professor at the Faculty of Medicine, Technion—Israel Institute of Technology, Haifa, where he also holds the Sylvia and Stanley Shirvan Chair in Cell and Tissue Regeneration Research. He obtained an M.D. degree from the Hebrew University in Jerusalem and a D.Sc. in Physiology from the Technion. While formerly an active researcher of assisted reproductive technnologies, his efforts have focused mainly on stem cells since his involvement in the isolation of the first human embryonic stem cell (hESC) line in Wisconsin in 1998. Subsequent to that breakthrough discovery, he established the first hESC laboratory in Israel and his team derived the first hESC lines in Israel in 2000. At that time he also engaged in collaborations in the USA and Europe for the distribution of hESC lines, and assisted numerous research groups in establishing stem cell laboratories, some of which have become leading research centers. Joseph Itskovitz-Eldor has participated in numerous scientific and ethical forums worldwide. Between 2007 and 2009 he served as president of the Israel Stem Cell Society.

M. Amit and J. Itskovitz-Eldor, *Atlas of Human Pluripotent Stem Cells: Derivation and Culturing*, Stem Cell Biology and Regenerative Medicine, DOI 10.1007/978-1-61779-548-0, © Springer Science+Business Media, LLC 2012

He currently devotes his research efforts toward developing novel technologies for the derivation and culture of hESC for research and clinical applications, and exploring the potential of these cells to differentiate into various lineages. With over 300 scientific publications in the fields of fetal physiology, assisted reproductive technologies, and human embryonic stem cells, Joseph Itskovitz-Eldor is an internationally acclaimed pioneer and a major contributor to the fields of reproductive medicine and stem cell research.

Index

MIX
Papier aus verantwortungsvollen Quellen
Paper from responsible sources
FSC® C105338

If you have any concerns about our products,
you can contact us on
ProductSafety@springernature.com

In case Publisher is established outside the EU,
the EU authorized representative is:
Springer Nature Customer Service Center GmbH
Europaplatz 3, 69115 Heidelberg, Germany

Printed by Libri Plureos GmbH
in Hamburg, Germany